IchigoJam My Reference

Preliminary Version - but still very useful

IchigoJam LPC1114 Board

USBtoTTL and Breadboard

Raspberry PI

```
2000 REM   IchigoJam Chip  /  CN5
2001 REM   1 VID1    KBD1 28  1 5V
2002 REM   2 VID2     SCL 27  2 SDA
2003 REM   3 IN1     KBD2 26  3 SCL
2004 REM   4 IN2    SOUND 25  4 3V
2005 REM   5 IN3      ISP 24  5 GND
2006 REM   6 IN4    RESET 23
2007 REM   7 VCC      GND 22  / USB
2008 REM   8 GND      VCC 21  1 GND
2009 REM   9 OUT1      X1 20  2 RXD
2010 REM  10 OUT2      X2 19  3 TXD
2011 REM  11 OUT3    OUT5 18  4 5V
2012 REM  12 OUT4    OUT6 17  5 3V3
2013 REM  13 BTN      TXD 16  6 DTR
2014 REM  14 LED      RXD 15
```

ABS	ANA	ASC	BEEP	BIN$	BPS	BTN	CHR$
CLEAR	CLK	CLO	CLP	CLS	CLT	CLV	CONT
COPY	DEC$	ELSE	END	FILE	FILES	FOR	FREE
GOSUB	GOTO	GSB	HELP	HEX$	I2CR	I2CW	IF
IN	INKEY	INPUT	IOT.IN	IOT.OUT	KBD	LANG	LC
LED	LEN	LET	LINE	LIST	LOAD	LOCATE	LRUN
NEW	NEXT	OUT	PEEK	PLAY	POKE	PRINT	PWM
REM	RENUM	RESET	RETURN	RND	RTN	RUN	SAVE
SCR	SCROLL	SLEEP	SOUND	SRND	STOP	STR$	SWITCH
TEMPO	THEN	TICK	UART	USR	VER	VIDEO	VPEEK
WAIT							

IchigoJam Instructions JPIN ExMark v11 July 2019

This is a translation from a Japanese documentation which has then be modified/adapted.

It started with a machine translated version of the Japanese reference.

Then other parts have been added as I find them suitable to understand IchigoJam.

This book has been delivered without any warranty.

No warranty can be given to the relevant software versions either.
This is supplied as is.

PERMISSION HAS BEEN GRANTED BY THE JAPANESE COPYRIGHT OWNERS TO USE THE MATERIAL IN THIS BOOK.

Version 11, July 2019

Copyright Juergen Pintaske, Exmark, UK

Introduction

IchigoJam – a system for Learning Programming for children and other interested parties; a project in Japan mainly.
The language used is BASIC. Is this a good starting point?
Well, I wonder, how many of the millions of programmers nowadays have started with BASIC. And BASIC was good enough for them then.
This project is aimed at the younger generation, and to get started. If they get into programming, at let's say the age of 10, it will take another 10 or more years if they do it as a profession – so the main aim is to attract the young generation into it, get some experience in how to write some little or more complex applications or examples. And nobody knows, which language might be the most important then when getting professional.
Rather than a PC, it is completely independent of any other hardware. It has as display – a low cost small monitor, as used in cars for rear view cameras.
The central chip is an NXP 1114 ARM processor. At the time of design, there was probably no low-cost chip with on-board USB available. So, a PS/2 keyboard was used – but here with a little interface that changes PS/2 to a USB connector.
Add a 5V power supply - for example as used for the Raspberry PI, and a 12V power supply for the little monitor, and programming can start. Or use a couple of batteries or rechargeable NMH cells and be independent.
And where is the Internet connection? There is none – and this actually adds to the system: children can be left on there own with no danger to go where they should not.
I had the opportunity to get one of these boards from Japan to the UK, and a new project started. Let's see what can be done to support this in English. There is a very active facebook group in Japanese and one in English – but translation works quite well – so no real issue.
But this was the issue for me, most of the data is in Japanese, and there are not enough code examples for me.
I contacted the relevant people in Japan and asked for permission to use some of the material for this book. As I walk through the different

instructions, I can as well document what I do – and add it to the material here. Aimed for others to try out and get comfortable.

This little board is very attractive for me, as I like minimal hardware solutions, as in other projects:

    The chip in the middle,
    Female headers as a way to connect to all of the pins
    The video output,
    The connector for power,
    the USB connector for the keyboard,
    A 5 pin expansion connector.

And do not forget that this board can be used "headless" – without monitor, keyboard and Power: connected via a USB-to-TTL to your PC and use your favourite Terminal Program – I started with Termite.

And there is a version for the microbit. Not tried yet, waiting for interface.

Raspberry PI is another option to run this IchigoJam BASIC software – just download, install the IchigoJam software and you have your new system. Most of the BASIC instructions work on the RPI version. I tried as well the little monitor and it worked. Hint: in UK swap " and @ usage on keyboard.

IchigoJam "PC-based" is another option to get started. Completely web-based, but then unfortunately no IO pins for external hardware, as I like it. I hope there will be a version soon, where USBtoTTL plus the actual board can be used for this. Or a serial data stream that can be used to drive an Arduino.

I hope this book **IchigoJam – My Reference** helps to spread the message about this project and can be used as a printed English documentation – easily available everywhere via amazon. I actually had to pay Import Tax in the UK for the delivery from Japan.

At the end of the book you find a list of links that I found. There are just a few. There might be better ones to be added later.

To be expanded in the next version.

Contents

Introduction	3
Contents	5
List of instructions	7
More detail per instruction	11
Commands	131
Constants	133
Functions	134
Statements	136
Testing the instructions	138
Link list part 1	139
Some program code preparation	141
IchigoJam instructions as block	146
Links	147
Olimex LPC1114 Board	148
LPC1114 pin functions	151
IchigoJam BASIC Refernce 1.3	152
Current list of examples	163
Fukuno-San's latest list 1.4 - comparison list to 1.3	172

There are many different versions of IchigoJam and as well different implementations (RPI and Web), but to make this book easier to read, most of this compatibility information has been left out.

The links at the end lead to other documentation where this additional information can be found. And as these products are always improved / updated / upgraded – a web based search is the best – so this book is actually just a starting position.

First here now as a preliminary version.
And only covering version 1.3.

This book is based on what I had acccess to and I used for the examples.

The web-based version has not been tried yet, and as such not been included here.

The two main links that were used as starting point for this book:

The original:
IchigoJam-BASIC repo by BALLOON a.k.a. Fu-sen. (IchigoJam Recipe) and
IchigoJam BASIC command list command reference (English) (-https://ichigojam.github.io/RPi/) (https://ichigojam.net/)

Then done by Paul Wratt as an automatic translation – so rather difficult to understand – but a lot better to understand than the original in Japanese for many.

https://github.com/paulwratt/IchigoJam-BASIC-english

My List of IchigoJam BASIC Instructions Ref 1.3

1	ABS	get the absolute value of a number	11
2	ANA	get the analog value of an input	12
3	ASC	return character code as number	14
4	BEEP	make a sound	15
5	BIN$	convert decimal number to binary	16
6	BPS	define Bits Per Second for serial comms	17
7	BTN	get the value of the button input status	18
8	CHR$	return character of character code	19
9	CLEAR	initialize variables and arrays	20
10	CLK	clear keyboard buffer	21
11	CLO	initialize Inputs and Outputs	22
12	CLP	initialize character font patterns	23
13	CLS	clear the screen	24
14	CLT	clear timer	25
15	CLV	clear variables and arrays	26
16	CONT	restart program execution	27
17	COPY	copy contents from A to B	28
18	DEC$	display number of aligned digits	29
19	ELSE	part of the IF ELSE THEN decision tree	30
20	END	end the running program, back to OK	32
21	FILE	get latest accessed file number	33

22 FILES	list the stored programs	34
23 FOR	part of the FOR NEXT loop	37
24 FREE	show how much memory is still free	39
25 GOSUB	go to a subroutine at line xx	40
26 GOTO	to another line rather than the next one	41
27 GSB	same as GOSUB	42
28 HELP	show the memory map	43
29 HEX$	convert decimal number to hexadecimal	45
30 I2CR	external extention via I2C interface READ	46
31 I2CW	external extension via I2C interface WRITE	48
32 IF	part of the IF ELSE THEN decision tree	50
33 IN	input logic status of defined pins	52
34 INKEY	return character pressed on keyboard	54
35 INPUT	prompt for input and put into variable	56
36 Iot.IN	receive serial information from IoT	58
37 Iot.OUT	output serial informatio to IoT	59
38 KBD	change the selected keyboard type	60
39 LANG	show selected language	62
40 LC	same as LOCATE	63
41 LED	control the on-board LED 0=off, 1=on	65
42 LEN	return the length of a character string	67
43 LET	assign a variable and set the value	68
44 LINE	return the line being executed	70

45 LIST	list program currently active in memory	71
46 LOAD	load a program into the active memory	72
47 LOCATE	locate the cursor at x.y on the screen	74
48 LRUN	load program in active memory and start	76
49 NEW	clear the active memory and show OK	78
50 NEXT	repeat loop specified via FOR	79
51 OUT	set selectedoutput pin to a digital value	81
52 PEEK	get data from a defined memory location	83
53 PLAY	play a sequence of notes	85
54 POKE	write a byte to a named memory location	87
55 PRINT	print a number or a string of charaters	89
56 PWM	Pulse Width Modulation Output at a pin	91
57 REM	text or code after REM is ignored to EOL	92
58 RENUM	renumber line number and step size	93
59 RESET	software reset, like switching off/on again	94
60 RETURN	at end of a subroutine back to next line	95
61 RND	get a random number between 0 and x-1	96
62 RTN	same as RETURN	97
63 RUN	start to execute a program	98
64 SAVE	save the program in active memory	99
65 SCR	same as SCROLL	101
66 SCROLL	scroll lines displayed up/down/left/right	103
67 SLEEP	put processor into low power mode	104

68 SOUND	make a sound	106
69 SRND	set random number seed	107
70 STOP	stop program execution	108
71 STR$	output a string of characters	109
72 SWITCH	switch between video and LCD display	110
73 TEMPO	set speed of notes played	111
74 THEN	part of the IF THEN ELSE decision tree	112
75 TICK	get internal counter value, steps of 1/60s	114
76 UART	set up the serial communication channel	116
77 USR	user function, for machine language	118
78 VER	get the version of installed IchigoJam SW	121
79 VIDEO	control the video output	124
80 VPEEK	peek into video memory, get byte there	127
81 WAIT	stop program execution, in 1/60 seconds	129

82 commands	List of commands	131
83 constants	List of constants	133
84 functions	List of functions	134
85 statements	List of statements	136

ABS

Function

Return the absolute value of a number.

Format

```
ABS(<value>)
```

Example

```
A = ABS(15)
I = ABS(A)
```

```
10 REM ABS() ABSOLUTE OF NUMBER
20 PRINT " ABS OF A=20 AND C=-20 "
30 LET A= 20
40 LET B = ABS(A)
50 PRINT B
60 LET C=-20
70 LET D = ABS(C))
80 PRINT D
```

Comment

Changes all numbers to positive numbers.
When <value> is a negative value (minus),
then change it to a positive value (plus) and return it.
Values greater than 0 (positive numbers) will be returned without a conversion.

ANA   2

Function

Returns the analog voltage at the IN2 and BTN terminals.

Format

```
ANA([<value>])
```

Example

```
A = ANA()
I = ANA(2)
```

1234567890123456789123456789012
(This line represents the 32 character width possible to display on screen, and is a placeholder to be seen in all further instructions;
this is where the examples will be added, but as this is rather time consuming, here first without the additional examples.
Some examples are already in the back, kindly supplied for this project, but have to be tried – and then addded.)

Comment

The resolution of the internal A to D converter is 10 Bit, so the returned value is between 0 to 1023
This is corresponding to the voltage of 0 V to 3.3 V – the power supply voltage of the chip.
1 V becomes about 310.
<Value> designation corresponds to value at num

2: IN2,
5-8: IN5-8 (OUT1-4)
0,9: BTN
0: omitted then BTN as well

How to test: connect relevant pin via a resistor of for example 470R to GND will return 0, connect pin via the resistor to Vcc will return 1023.

One resistor to GND and Input and one resistor to VCC and the same input gives a voltage divider. Can be used with a little bit of calculation to measure unknown resistance.

A fun bit is to connnect an LDR to ground and the analog input, and an additional restor of I think 20kOhm to VCC. Covering the LDR will change the input voltage. Or switching the light in the room on and off. Can be used for counting.

IchigoJam BASIC RPi unfortunately always returns 0 as there is no internal A/D converter, needs an additional external A/D Board.

ASC 3

Function

Returns the character code number corresponding to the character.

Format

```
ASC("<character>")
ASC(<number of addresses>)
```

Example

```
PRINT ASC("A")
A = ASC("B")
B = ASC(C)   (1.2 beta 3 ~)
```

123456789012345678912345678 9012

Comment

You can specify <address>. This is usually a variable.
Returns the character code at the address referenced by the variable.
The variable is supposed to be an address indicating a character string.

BEEP 4

Function

It will issue BEEP.

Format

BEEP [<period>, <length>]

Example

BEEP

123456789012345678912345678 9012

Comment

<period> is 1 to 255, <length> is for example 60 and it is then 1 second. This can be omitted.

In direct mode, playback stops when another command is executed.

Connect a piezoelectric sounder to SOUND and GND to produce the sound.
In IchigoJam shipped since May 2014 - January 2015
The SOUND terminal is described as EX2.

IchigoJam BASIC RPi supports it from 1.2b17.
Connect the piezoelectric sounder to physical PIN 29 and GND.

BIN$ 5

Function

Convert decimal number to binary notation.

Format

BIN$(<value> [, <number of digits>])

Example

```
BIN$(255)      → 11111111 will be returned
BIN$(-1,16)    → 1111111111111111 will be returned
```

123456789012345678912345678 9012

Comment

When <number of digits> is omitted, it outputs with the minimum number of digits.
If the number of digits to be output is small, fills the head with 0.

BIN$ will be used for PRINT.

BPS   6

Function

Sets the communication speed of serial communication.

Format

```
BPS [<communication speed>]
BPS [<communication speed>] [, <I2C communication speed>]
```

Example

```
BPS 9600
BPS 0
```

123456789012345678912345678 9012

Comment

Set the value of <communication speed> in bps – bits per second.
Other than that, the following values and abbreviations are valid.
    Abbreviation
        0 115200 bps
        -1 57600 bps
        -2 38400 bps
A setting of -100 or less is supported.
For example, it is set to 230400 bps at -2304.
<I2C communication speed> can be set for 1.2 b 56 or later and 1.2 b 18 on RPi or later. The value is in kHz, default is 400.
Before 1.2.3, 400 kHz was fixed.

BTN   7

Function

It returns the state of the button switch if pushed or not.

Format

```
BTN([<value>])
```

Example

```
PRINT BTN()
I = BTN(SPACE)
```

123456789012345678912345678 9012

Comment

Returns a 1 if the button is pressed, 0 if not pressed.

BTN() and BTN(0) are valid only for IchigoJam BASIC where the button exists on the board.

In most later versions, the BTN terminal is assigned to IN9.
See also IN.

CHR$ 8

Function

Returns the character corresponding to the character code number.

Format

```
CHR$(<character code>)
CHR$(<character code>, <character code>, ...)
```

Example

```
PRINT CHR$(65)
PRINT CHR$(1,0,1,1,1,0,1)
```

123456789012345678912345678901 2

Comment

Mostly CHR$ will be used with PRINT.

Multiple character codes can be inserted with the delimiter ",", and all characters can be output.

CLEAR 9

Function

Initialize the defined variables and arrays to 0.

Format

```
CLEAR
CLV     as a shorte command
```

Example

```
CLEAR
CLV

123456789012345678912345678901 2
```

Comment

Variables and arrays can now be inherited.
It is also possible to set and inherit variables and arrays in direct mode.
Therefore, initialization of variables and arrays is done by using CLEAR.

CLV is an alias for CLEAR.

CLK 10

Function

Clear keyboard buffer and key state.

Format

CLK

Example

CLK

123456789012345678912345678 9012

Comment

When several keys are input, (Especially when the key repeat occurs with keeping the key pressed).
Since the input is stored in the key buffer it may adversely affect input acceptance such as INKEY () or INPUT.
By clearing the buffer with CLK, we can deal with this issue.

CLO   11

Function

Initialize the defined pins to input / output state.

Format

```
CLO
```

Example

```
CLO
```

1234567890123456789123456789012

Comment

Set the input and Output Pins to the initial status.

CLP 12

Function

Initialize the character font rewritten with POKE.

Format

CLP

Example

CLP

123456789012345678912345678 9012

Comment

If character codes 224 (#E0) to 255 (#FF) were rewritten with POKE, returns the fonts to the initial state.

CLS   13

Function

Clear the screen.

Format

`CLS`

Example

`CLS`

`123456789012345678912345 6789012`

Comment

Erase all characters displayed on the screen.

When connected with serial, screen clear will work with the corresponding application.
You can change this behavior via the UART.

Only the range of the screen area is cleared.
The area outside the screen does not change.
Therefore, unused areas can be used for another purpose.

CLT   14

Function

Reset elapsed time counter.

Format

CLT

Example

CLT

123456789012345678912345678 9012

Comment

The value returned by TICK is reset to 0.
Since v0.9.8, time since execution of RUN is not reset to.
Depending on the program, it is necessary to rest the Tick counter with CLT to 0.

CLV   15

Function

Initialize variables and arrays.

Format

CLV
CLEAR

Example

CLV
CLEAR

123456789012345678912345678 9012

Comment

Up to 0.9.3 variables and arrays were initialized at the stage of RUN,
From 0.9.4 they ares now maintained without being initialized.
As a result, when another program is executed by LRUN from the program, then
Variables and arrays can now be inherited.
It is also possible to set and inherit variables and arrays in direct mode.
Therefore, initialization of variables and arrays is done using CLV if needed.

CLEAR can be used instead of CLV.

CONT   16

Function

Restart the program execution from the line where the program execution stopped.

Format

CONT

Example

CONT

123456789012345678912345678 9012

Comment

A program is stopped with the Esc key,
Or with the STOP command
Then program execution resumes.
CONT is not a continuation from STOP.

Please be careful to execute from the beginning of the stopped line.

COPY   17

Function

Copy the contents of the virtual memory area.

Format

```
COPY <area destination>, <area source>, <number of transfers>
```

Example

```
COPY #880,#800,5
```

1234567890123456789123456789012

Comment

From **<source>** of virtual memory area to **<destination>**
<Number of forwarding> .
Performs byte transfer.
<Source> → <destination>, <destination> + 1 → <destination> + 1, ...

For example, fill the full screen with the strawberry mark in the following cases.

CLP: POKE #900,255: COPY #901,#900,767

From 1.2 beta 21, you can use minus value for <transfer count>.
If you set <transfer count> to minus, copy from the end.
1.2 beta 23 or later <area source> <area destination> specifies the end of the area.

DEC$ 18

Function

Displays the number of digits aligned.

Format

```
DEC$(<value>, <number of digits>)
```

Example

```
DEC$(123,5)
DEC$(I,3)
```

```
123456789012345678912345678901 2
```

Comment

Display <value> in <number of digits>.
Blanks are filled for missing digits.
When you extend the number of digits, it displays the low order. (DEC$(123,2) → 23)

IF~THEN(~ELSE)

Function

Perform conditional branching.

Format

```
IF <condition value> [THEN] <command>
IF <condition value> [THEN] <command> [ELSE <command>]
```

Example

```
IF A=B THEN LED 1
IF A=B LED 1

123456789012345678912345678901 2
```

Comment

Execute <command> when <condition value> is correct (when it is not 0).
<condition value> corresponds to multiple conditions by AND and OR from 0.9.4.
Up to 0.9.3 substitute with * and + or & and |. (Condition values before and after are delimited by () pair)

Multiple executions of <command> are possible, separated by ":".
It will cause the command within that line to continue.

Although THEN can be omitted, the line number becomes an error immediately after THEN.

To skip, please use GOTO line number or THEN GOTO line number.

When <condition value> is wrong (in case of 0),
you can execute the command immediately after ELSE.
It is optional.
ELSE is supported from 0.9.4.

END   20

Function

End program execution.

Format

END

Example

END

123456789012345678912345678 9012

Comment

With this command END in the program, we end the execution of the program.
Even if it is used in direct execution mode, it will not result in an error.

Only with 1.1 beta 6, Syntax Error will be generated due to malfunction.

FILE

Function

Get the latest accessed file number.

Format

FILE()

Example

PRINT FILE()

123456789012345678912345 6789012

Comment

Returns the most recently accessed file number that used SAVE · LOAD · LRUN.
If file access has not been performed from power on,
a -1 is returned.

FILES   22

Function

The list of stored programs is displayed.

Format

```
FILES [[<start file number>,] <end file number>] (1.1.0 -)
FILES [<Maximum number of files>] (~ 1.0.2)
```

Example

```
FILES
FILES 0, 163 (1.1 beta ~)
FILES 68 (~ 1.0.2)
```

123456789012345678912345678 9012

Comment

In IchigoJam main unit, including EEPROM and when external EEPROM is connected
List the programs.

After the existing program number, it displays the characters in the first line of the program.
'If you put a comment in this line it will be displayed so it is convenient.

In the case of external EEPROM of 512K and 1024K bits,
except for some EEPROMs, they are displayed completely.
For the 32K to 256K bits the same program.

Please be careful to display with multiple program numbers.
Because there is a difference in the access of I2C for EEPROM of 16 K bit or less, they can not be used for saving programs.
From 1.1 beta (including 1.0.2 beta 11),
the file is displayed in the range from <start file number> to <end file number>.
For FILES only, display only the main body, FILES 0 will show everything including EEPROM.
, You can also specify only <end file number> without removing.
FILES 110 will be the same as FILES 0, 110.

For 1.0.1 · 1.0.2 beta,
Specify the maximum number of files that can be displayed when all files exist.
For 8, it is 0 to 3 and 100 to 103, it is a maximum of 8 files.
Specifically, it is as follows.

```
   8    = 103 (32 Kbits)
   9~12 = 107 (64 Kbits)
  20    = 115 (128 Kbits)
  36    = 131 (256 Kbits)
  68    = Up to 163 (512 Kbits)
   0    = Show all
```

You can also display 1 page, 2 page, 3 page as 22, 44, 66.

If there is no file, it will be hidden,
Since this is also counted, please be aware that it may be displayed less than the specified number of files.

In the range designation from 1.1 beta, the best display is made as follows.
(When 512K or less is assumed for A2 = GND connection)

```
Chip only    = FILES or FILES 0, 3
 32 Kbits    = FILES 103 or FILES 0, 103
 64 Kbits    = FILES 107 or 0, 107
128 Kbits    = FILES 115 or 0, 115
256 Kbits    = FILES 131 or 0, 131
512 Kbits    = FILES 163 or 0, 163
All Display  = FILES 0 or FILES 0, 227
```

FILES 0, 0 are treated the same as FILES 0 until 1.2.3, and all are displayed.
1.2.4 · 1.2b56 · 1.2b20 From RPi,
FILES 0 and 20 only show program number 0.

For IoT compatible version, only 0 is displayed in case of FILES from 1.2b59 IoT.

IchigoJam PC tries to display up to 227,
Actually, you can only use 0 to 3.

IchigoJam BASIC RPi is 0 to 3 and 100 to 227 to 1.2b15 RPi and 1.2.4 RPi.

(micro) Refer to the file in the files directory of the SD memory card.
It became 0 to 227 from 1.2b16 RPi and 1.2.5 RPi.
0 to 99 are SD memory cards, 100 to 227 refer to EEPROM.
If omitted, the same applies to FILES 0 and 3 up to 1.2.4 RPi and 1.2b19 RPi,
For 1.2b20 RPi and 1.2.5 RPi and later, FILES 0 and 20 are the same.

FOR~NEXT   23

Function

Repeat loop with the range of values specified by FOR.

Format

```
FOR   <variable>  =  <start  value>  TO  <end  value>  [STEP
<difference>]
FOR   <variable>,  <start  value>  TO  <end  value>  [STEP
<difference>]  (1.0.0 beta 3 ~)
:
NEXT
```

Example

```
FOR I=1 TO 11 STEP 2
PRINT I
NEXT

123456789012345678912345678901 2
```

Comment

From <start value> to <end value> in FOR,
execute the commands etc. that are defined up to NEXT.
A value can be entered via a <variable>, and this variable can be used.
You can change the number added by setting STEP <difference>.
When omitting it, it will add in steps of one.
Four loops are possible. (6 more than 1.2 beta 10)

No variable is added to NEXT.

From 1.0.0 beta 3, you can also use "," instead of "=" in FOR.

FREE 24

Function

Returns available memory space still unused in the program.

Format

FREE ()

Example

PRINT FREE ()

123456789012345678912345678 9012

Comment

Returns the free space for the currently active program.
The initial state of the memory available where nothing has been entered after NEW is 1022 or 1024.

GOSUB   25

Function

Transfer processing from the main program to a subroutine.

Format

```
GOSUB <line number>
or
GSB <line number> (1.1.0 beta 1 ~)
```

Example

```
GOSUB 100
GSB 100 (1.1.0 beta 1 ~)

123456789012345678912345678012
```

Comment

Memorize the next line number, and then transfer processing to the line specified by <line number>.

You return from the execution of a subroutine started by GOSUB to continue from the memorized line of the calling program with RETURN.

From 0.9.6 it is possible to use calculation formulas, variables etc for <line number>.

GOTO 26

Function

Specifies the line number to be executed next, rather than the next line

Format

```
GOTO <line number>
```

Example

```
GOTO 10
```

```
123456789012345678912345678 9012
```

Comment

Transfers processing to the line specified by <line number> rather that the next line.

From 0.9.5 it is possible to use calculation formulas, variables, etc. for <line number>.

GSB   27

Function

Transfers processing to subroutine.

Format

```
GSB <line number>  (1.1.0 beta 1 ~)
GOSUB <line number>
```

Example

```
GSB 100  (1.1.0 beta 1 ~)
GOSUB 100
```

123456789012345678912345678012

Comment

Transfers processing to the line specified by <line number>.
You return to the process for continuation with RETURN.

From 0.9.6 it is possible to use calculation formulas, variables etc for <line number>.

HELP   28

Function

Display the memory map.

Format

HELP

Example

HELP

123456789012345678912345678912

Comment

The following contents is displayed

```
MEMORY MAP
#000 - #6FF  CHAR PATTERN (ROM)
#700 - #7FF  CHAR PATTERN (RAM)
#800 - #8FF  VARS
#900 - #BFF  VRAM
#C00 - #FFF  PROGRAM
```

1.1 beta 8 to 15 are the contents of the following:

```
MEM MAP
#000 CHAR
```

```
#700  PCG
#800  VAR
#900  VRAM
#C00  PROGRAM
#1002 KEY
```

1.1 beta 16 ~ and 1.1.1 have the following contents.

```
MEM MAP
#000  CHAR
#700  PCG
#800  VAR
#900  VRAM
#C00  PROGRAM
```

1.2 has the following contents.

```
MEM MAP
#000  CHAR
#700  PCG
#800  VAR
#900  VRAM
#C00  LIST
```

HEX$ 29

Function

Convert a decimal number to a hexadecimal notation.

Format

```
HEX$(<value> [, <number of digits>])
```

Example

```
HEX$(255) → FF is returned
HEX$(-1, 4) -> FFFF will be returned
```

123456789012345678912345678 9012

Comment

When <number of digits> is omitted, it outputs with the minimum number of digits.
If the number of digits to be output is small, fill the head with 0.

Because IchigoJam can only handle variables and arrays with numbers, Use of HEX$ will be with PRINT.

I2CR   30

Function

It retrieves data from the device connected by the I2C serial bus, and store it in local memory.

Format

```
<return value> = I2CR(<I2C address>, <command start address>,
<command length>, <read memory start address>, <read memory
length>)
<return value> = I2CR(<I2C address>, [<command start address>,
<command length>,> <read memory starting address>, <read memory
length>) (1.2b42 ~)
```

Example

```
R = I2CR(`1010000,#700,2,#702,16)
R = I2CR(`1010000,#700,16)  (1.2 b 42 -)

123456789012345678912345678 9012
```

Comment

<I2C address> is the device to obtain information from via I2C.
Depending on the connection of the A0 to A2 terminals to the I2C device, it has a unique slope address. Specify this.
Values range from 0 to 127 (`0000 000 to `1111 111).
After adding the READ / WRITE bit READ to this and transmitting it,
It gets data received from the external device.

Please note that it does not include the READ / WRITE bit.

Next from <command start address> to <command length>
Execute the command.
Prepare the command beforehand with POKE.

After that, the data obtained from I2C is read from <read memory starting address>
Store with <reading memory length>.
Please obtain actual value with stored information by PEEK.

<command start address> <read memory starting address> is
IchigoJam's memory location.
<command length> <read memory length> is
Indicates the length in bytes.

By pressing the Esc key at 1.0.0 beta 11
It is now possible to interrupt reception.

From 1.2b42, with <command start address> and <command length>,
You can now omit it. Regardless of the command,
it is now possible to receive arbitrary data.

I2CW 31

Function

The values in local memory
are sent to the external device connected via the I2C serial bus.

Format

```
<return value> = I2CW(<I2C address>, <command start address>, <command length>, <write memory start address>, <write memory length>)

<return value> = I2CW(<I2C address>, [<command start address>, <command length>,] <write memory start address>, <write memory length>) (1.2b42 ~)
```

Example

```
R = I2CW(`1010000,#700,2,#702,16)
R = I2CW(`1010000,#700,16)  (1.2 b 42 -)

123456789012345678912345678 9012
```

Comment

<I2C address> is the device to send information to via I2C.
Depending on the connection of the A0 to A2 terminals of the I2C device,
it has a unique slope address. Specify this.
Values range from 0 to 127 (`0000 000 to `1111 111).
Add WRITE of READ/WRITE bit to this and transmit.
Please note that it does not include the READ/WRITE bit.

After that, the command at <command length> from <command start address>
From <writing memory starting address> to <writing memory length>
Send the data.

<command start address> <writing memory starting address>
is the memory location containing the value to pass from IchigoJam.
<command length> <write memory length> indicates the length in bytes.
We will send the data in I2CW after putting in the value in advance with POKE.

By pressing the Esc key at 1.0.0 beta 11
it is now possible to interrupt transmission and reception.

From 1.2b42, with <command start address> and <command length>,
you can now omit it.
This allows you to separate commands and data that
you can now send out.

IF~THEN(~ELSE)   32

Function

Perform conditional branching.

Format

```
IF <condition value> [THEN] <command>
IF <condition value> [THEN] <command> [ELSE <command>]
```

Example

```
IF A=B THEN LED 1
IF A=B LED 1
```

1234567890123456789123456789012

Comment

Execute <command> when <condition value> is correct (when it is not 0).

<condition value> corresponds to multiple conditions by AND and OR from 0.9.4.

Up to 0.9.3 substitute with * and + or & and |. (Condition values before and after are delimited by () pair)

Multiple executions of <command> are possible, separated by ":".
It will cause the command within that line to continue.

Although THEN can be omitted, the line number immediately after THEN becomes an error.
To skip, please use GOTO line number or THEN GOTO line number.

When <condition value> is wrong (in case of 0),
you can execute the command immediately after ELSE. It is optional.
ELSE is supported from 0.9.4.

IN   33

Function

Get the logical value at an input of IN terminal.

Format

`IN([<terminal>])`

Example

```
LET A,IN(1)
A=IN()
```

123456789012345678912345678

9012

Comment

When <terminal> is omitted, the state of IN1 - 4 is represented by 1, 2, 4, 8. In this case the return value is in hexadecimal from 0 to 15.

When attaching <terminal>, the range of <terminal> is 1 to 4, and the return value is 0 or 1.

Commands with <terminal> are valid after 0.8.7.
IN1 and 2 and 4 have pull-up resistance.
BTN has resistor R3 (1 MΩ).
Note that IN 3 has no resistance.

1.0.0 beta 3 ~ 1.1 Beta 4 treats BTN as IN5.
When <terminal> is omitted, it is represented by 16, and when adding <terminal> it is 5.

Assign the OUT1 to OUT4 terminals to IN5 to IN8.
<Terminal> When it is omitted, it becomes 16, 32, 64, 128, and when attaching <terminal> it becomes 5 to 8.
BTN is changed to IN9 handling, and when attaching <terminal> is 9.
1.1 beta 5 returns BTN operation with IN() in 32, there is an issue with the operation.

In 1.1 beta 6 and later, by specifying OUT 5, -1 and OUT 6, -1, assign the OUT 5 to OUT 6 terminals to IN 10 to IN 11.
<Terminal> When 512 and 1024 is omitted, it becomes 10 to 11 when attaching <terminal>.

It corresponds to the pull-up input by designation of OUT 5, -2 etc. from 1.2 beta 43.

SkyBerryJAM also has IN1 and IN4 buttons.
KumaJam also has an IN1 button.
Pressing these buttons will change the value of IN.
Unlike BTN, be careful about that 1, push to 0 after separating.

IchigoJam BASIC RPi has no board button in Raspberry Pi,
Instead of the BTN terminal, it is treated like an independent IN9 terminal.

IchigoJam web · IchigoJam Returns 0 as the IN terminal does not physically exist on PC.
Please be careful when considering the button with the IN terminal.

Depending on IN2 version, BTN and OUT1 ~ 4 (IN 5 ~ 8),
the state of the voltage can be obtained as 0 to 1024. (Analog input)
See ANA.

INKEY 34

Function

Returns the character being pressed on the keyboard.

Format

`INKEY()`

Example

`PRINT INKEY()`

`123456789012345678912345678012`

Comment

Returns the character code of the key being pressed.
If none is pressed, 0 is returned.

1.2 - If beta 19 gets CHR$(0) from serial,
It returns 256 (#100).

In other versions, the return value of INKEY$ returns the character itself.
In IchigoJam, variables are numbers, so they are in code.
If you want to use letters please use together with ASC.

`10 I=INKEY(): IF I<>ASC("0") GOTO 10`

For the example above, wait until 0 key is pressed (until you get 0 from serial).

The special codes are as follows:

```
10 · Enter (return)
28 · LEFT ← (left)
29 · RIGHT → (right)
30 · UP ↑ (upper)
31 · DOWN ↓ (bottom)
```

```
10 I = INKEY(): IF I<>10 GOTO 10
```

Wait in code above until Enter key is pressed.

Successive key inputs are stored in the key buffer and sequentially fetched by INKEY().

Auto-repeat is effective when long pressed.
For game operation purposes, this may result in unintended behavior.

The key buffer can be cleared using CLK. See CLK.
Also, by using BTN(<value>) instead of INKEY()
You can get the current key state unaffected by the key buffer. Please refer to BTN.

INPUT   35

Function

Prompt for input and put it in the variable.

Format

`INPUT ["<string>",] <variable>`

Example

`INPUT "ANS?", A`

`123456789012345678912345678901`

Comment

You will be prompted to enter a number until Enter (return) is entered.
The entered numerical value enters <variable>.
If there is <character string>, it will display and continue input.
If there is not, ? Is displayed.

If a character or symbol other than numbers is entered in the input, it will not result in an error.
After the letters, other than numbers in the head, are ignored (even if there are numbers).
If there are no digits in the head, 0 is returned if nothing is entered.

Normally you will enter from the keyboard, when serial connection is performed, input from serial is accepted.

From 1.2 beta 3 it is handled as a character string by enclosing with "",
Returns the starting address of the string.
Since this address is returned from the screen position #900 to #BFF,
Please note that you lose the character string when scrolling.

IoT.IN   36

Function

Receive data from sakura.io module.

Format

```
IoT.IN()
```

Example

```
A=IoT.IN
```

123456789012345678912345678 9012

Comment

Relates to sakura.io module.
Using the data reception command (Q=#30),
get the data at the head of the receive queue.
The channel of the response data is ignored and all data is received,
Returns the value with D(0) as the lower byte and D(1) as the upper byte.

IoT version executes IoT.OUT
Virtual memory area is #800 or higher.
It is used for data transmission command generation.
Since this area is used in an array,
please note that part of the array will be damaged.

IoT.OUT

37

Function

Send data to sakura.io module.

Format

```
IoT.OUT <value>
```

Example

```
IoT.OUT 1
```

123456789012345678912345678912

Comment

For sakura.io module.
Using the data transmission command (Q=#20)
Send the value and add it to the send queue.
The channel at transmission is fixed at #01,
type is treated as an unsigned 64 bit integer (#4C).

IoT version executes IoT.OUT
Virtual memory area #800 or higher.
It is used for data transmission command generation.
Since this area is used in an array,
Please note that part of the array will be damaged.

KBD

Function

Change the keyboard layout.

Format

`KBD <keyboard array>`

Example

`KBD 1`

`123456789012345678912345678 9012`

Comment

On IchigoJam BASIC RPi it is a uspecial command.
In IchigoJam BASIC by LPC 1114, if you want to change the keyboard layout, update the firmware.

<keyboard array> works as follows.

   0 English (US) keyboard
   1 Japanese (JIS) keyboard

It is also possible to set it with the file keymap.txt to be put onto the SD card.

Also, kapmap.txt will be overwritten via the KBD command.

Options are as follows:

　us English (US) keyboard
　jp Japanese (JIS) keyboard

This command is implemented exclusively for IchigoJam BASIC RPi. IchigoJam: On the actual board, change the keyboard layout with a firmware update.

LANG   39

Function

Returns the language type.

Format

LANG()

Example

```
L=LANG()
PRINT LANG()
```

1234567890123456789123456789012

Comment

The values to return are as follows:

    1 Katakana and Romaji input
    2 Cyrillic letter with Mongolian arrangement
      (IchigoJam mn)
    3 Vietnamese (IchigoJam vi)
    4 AZERTY sequence (IchigoJam fr)
    6 Chinese Bopomofo Zhuyin Symbol (IchigoJam bp)

IchigoJam BASIC RPi and IchigoJam PC:
Since only Kana character and Roman character input is provided, it is fixed to 1.

LC - short version of LOCATE   40

Function

Specify the position of the character to be displayed.

Format

```
LOCATE <abscissa>, <ordinate>
LC <abscissa>, <ordinate>  (0.8.9 ~)
```

Example

```
LOCATE 3,3
LC 5,5

123456789012345678912345678 9012
```

Comment

<abscissa> = <x-axis>
<ordinate> = <y-axis>
<vertical coordinate> = <ordinate> = <y-axis>

The display range for <abscissa> was 0 to 35, <vertical coordinate> is 0 to 26.

From 0.9.9, <abscissa> is 0 to 31, <vertical coordinate> is 0 to 23,

IchigoJam BIG published at 1.2.0
<abscissa> is 0 to 15, and <vertical coordinate> is 0 to 11.

However, even if this value is exceeded, it will not result in an error, and move the cursor to the rightmost or bottommost point.

LC can be used instead of LOCATE.

By setting the <ordinate> to -1, it does not output to the video input screen, but
it outputs only to the serial port. From 0.9.4.

LED   41

Function

Turn the LED on the board on and off.

Format

```
LED <number>
```

Example

```
LED 1
LED 0
```

```
123456789012345678912345678912
```

Comment

IchigoJam illuminates or turns off the lamp attached to the main unit.
IchigoJam PC turns on / off the keyboard LED.
IchigoJam BASIC RPi turns the ACT LED on / off.
<Number> lights up with any value other than 0, LED0 turns off.
For 1.1 beta or later and 1.0.2 beta 3 to 11
For the LED OUT7 can be used.
For example: OUT 7, n can operate the same as LED n.
See OUT.

Of the five lamps in SkyBerryJAM, the leftmost changes with the LED command.

The other four lamps are equivalent to OUT1 to 4 terminals. They change with the OUT command.

Since LED and OUT7 are part of the OUT, they can be controlled all at once via the OUT command.

In IchigoJam BASIC RPi, OUT7 and LED are handled separately.

LEN

42

Function

Returns the length of the selected string.

Format

```
LEN(<start address>)
```

Example

```
PRINT LEN(A)
```

123456789012345678912345678 9012

Comment

Returns the length of the character string specified by <start address>.
Normally this is a variable, but you can also specify A+1.
From this address up to where " or CHR$(0) is located.
Returns the number of bytes.

LET 43

Function

Put values into variables / arrays. And define theem if they do not exist yet.

Format

```
LET <variable>, <number>
<variable> = <number>
```

```
LET [<array>], <number>
[<array>] = <number>
```

```
LET [<array>], <number>, <number>, ...   (0.9.4 -)
<variable> = "<string>"   (1.2 beta 2 ~)
```

```
LET <variable>, "<string>"   (1.2 beta 2 ~)
LET [<array>], "<string>", "<string>", ...   (1.2 beta 2 ~)
```

Example

```
LET A,1
A=1
LET [0],0
LET [1],1,2,3,4,5
A="ABC"
LET B="BCD"
```

123456789012345678912345678901

Comment

<variable> is an alphabetic character. You can put values into the array at the same time.
Because IchigoJam's variables are numeric and integer only,
<number> is an integer. The range is from -32768 to 32767.

The delimiter in the notation with LET is ","
Please note that it is different from "=" if omitted.

When putting data into an array with LET, write multiple <number>
It can be put in multiple arrays. It corresponds from 0.9.4.

In case of LET [1], 1, 2, 3, 4, 5, values 1 to 5 are entered in [1] to [5] respectively.

It corresponds to the character string from 1.2 beta 2.
<variable> contains the start address of the address containing the character string.
Normally this will return program area #C00 - #FFF.

LINE   44

Function

Get the line number being executed at the moment.

Format

LINE()

Example

PRINT LINE()
L=LINE()

123456789012345678912345678912

Comment

Returns the line number being executed.
If the program is not running, 0 is returned.
Used for example for debugging purposes to control the program structure.

LIST

45

Function

List the current active program.

Format

```
LIST [[<start line>,] <end line>]
```

Example

```
LIST
LIST 100
LIST 10,300

123456789012345678912345678 9012
```

Comment

If <start line> and <end line> are omitted, all program lines are displayed.

If you specify only one line number without appending, then you have specified <end line>
The program is displayed from the beginning.

From 1.2 beta 43 the display target is set to line number 16384.
1.2 When beta 44 sets <end line> to 0,
If 16385 or more is entered in <start line>, it displays up to the last line including 16385 onwards.

LOAD   46

Function

Load a program into memory.

Format

```
LOAD [<program number>]
```

Example

```
LOAD
LOAD 20
```

123456789012345678912345678 9012

Comment

Any programs currently loaded into IchigoJam are erased.
It is changed to the program loaded via LOAD.

The value of <program number> is 0 to 2 or 0 to 3 and
It is 100 to 226 or 100 to 227 depending on version.

The IoT compatible version is only 1.2b59 IoT and the program is only saved as 0.
If omitting it immediately after startup, reads 0.

From 0.9.3, save to external EEPROM. 100 to 226 is valid.

Values from 100 to 227 are valid from 0.9.9 RC.
Valid values vary depending on the EEPROM capacity and wiring.
Wrong program number returns File Error.

EEPROM can be used with 32K to 1024Kbits,
When using 1024 Kbits frm 0.9.9 and
Contents can be completely used except some EEPROM types.
For the 32K to 256K bits the same program
Please be careful to refer with multiple program numbers.
Because there is a difference in the access of I2C for EEPROM of 16K bit or less,
It can not be used for saving programs.

IchigoJam BASIC RPi it is 0 to 3 and 100 to 227 for 1.2 b 15 RPi · 1.2.4 RPi.
(micro) Read from the file in the files directory of the SD memory card.
It became 0 to 227 from 1.2b16RPi · 1.2.5RPi.
0 to 99 are SD memory cards, 100 to 227 are read from EEPROM.

If there is no program in the related program number (SAVE with NEW),
it will not result in an error, it will show OK display.

You can load and execute immediately after switching on power. See LRUN.

Technically, 1 Kbyte is the program area of the main memory and EEPROM
It is copied to virtual address area #C00 ~ #FFF.
For this reason, POKE puts in a vacant area of the program (area behind #FFF) with a value.
Save this value including SAVE,
After LOAD or LRUN this value can be retrieved with PEEK.

LOCATE 47

Function

Specify the position of the character on screen where it will be displayed.

Format

```
LOCATE <abscissa>, <ordinate>
LC <abscissa>, <ordinate> (0.8.9 ~)
```

Example

```
LOCATE 3,3
LC 5,5 (0.8.9 ~)
```

123456789012345678912345678912

Comment

<abscissa> = <x-axis>
<ordinate> = <y-axis>
<vertical coordinate> = <ordinate> = <y-axis>

The display range is version 0.9.8, <abscissa> is 0 to 35, <vertical coordinate> is 0 to 26,

After 0.9.9, <abscissa> is 0 to 31, <vertical coordinate> is 0 to 23,
For IchigoJam BIG published at 1.2.0
<abscissa> is 0 to 15, and <vertical coordinate> is 0 to 11.

However, even if it exceeds this value it will not result in an error, but move the cursor to the rightmost or bottommost point.

LC can be used instead of LOCATE.

Serial connection is from version 1.1.0 or later,
LOCATE can be reflected by using the corresponding application.

By setting the <ordinate> to -1, it does not output to the video input screen,
It outputs only to the serial port from 0.9.4.

LRUN 48

Function

Load the program and execute it automatically.

Format

```
LRUN [<program number>]
LRUN [<program number>] [, <line number>]
```

Example

```
LRUN 1
LRUN 3,200 (1.0.0 beta 4 ~)
```

123456789012345678912345678 9012

Comment

Programs that are currently loaded into IchigoJam memory are erased. It is changed to the program loaded with LRUN.

Variables and arrays are initialized up to version 0.9.3.
It will not be initialized after 0.9.4.
Please use CLV or LET.

The value of <program number> is 0 to 3 and 100 to 227.
IchigoJam main body corresponds to 0.9.9 from 0 to 2, 1.0.0 to 0 to 3.
When omitted, it is 0 up to 0.9.8,

It becomes the program number which had been accessed latest from 0.9.9.
The IoT compatible version is only for version 1.2b59 IoT and the program is only saved to 0.
If it is omitted immediately after startup, 0 is executed.

From 0.9.3, save to external EEPROM, 100 to 227 is valid.
Valid values vary depending on the EEPROM capacity and wiring.
A program number that can not be used sends File Error.

EEPROM can be used with 32K to 1024Kbits.
When using 1024Kbits at 0.9.9
And contents can be completely used except for some EEPROM.
For the 32K to 256Kbits for the same program, please be careful to refer with multiple program numbers.
Because there is a difference in the access of I2C for EEPROM of 16K bit or less, it can not be used for saving programs.

If there is no program in the program number (SAVE with NEW), it will not result in an error, it will be an OK display.

If you only want to load programs, use LOAD.

LRUN can also be included in a program.
This allows you to navigate through multiple programs, and as such you can also make long programs than the 1024 bytes long.

<line number> can be specified from 1.0.0 beta4.
This allows execution from a specific line in the newly loaded program.

Technically, it uses the 1 Kbyte area of the main body and EEPROM.
It is copied into virtual address area #C00 to #FFF.
For this reason, POKE puts in a vacant area of the program (area behind #FFF) with a value
Save this value including SAVE,
After LOAD and LRUN this value can be retrieved with PEEK.

NEW   49

Function

Erase the current program in memory.

Format

NEW

[Example]

NEW

123456789012345678912345678 9012

Comment

Delete all currently loded programs.
Only programs that can be displayed with LIST are erased.
Using the SAVE program does not erase the currently loaded program.

FOR~NEXT 50

Function

Repeat loop with a range of values specified via FOR.

Format

```
FOR <variable> = <start value> TO <end value> [STEP <difference>]

FOR <variable>, <start value> TO <end value> [STEP <difference>] (1.0.0 beta 3 ~)
:
NEXT
```

Example

```
FOR I=1 TO 11 STEP 2
PRINT I
NEXT
```

123456789012345678912345678901 2

Comment

From <start value> to <end value> in FOR,
execute existing commands etc. up to NEXT.
A value can be entered in <variable>, and this variable can be used.
You can change the number added by setting STEP <difference>.
When omitting it, I will be a step size 1.
Four loops are possible.

No variable is added to NEXT.

From 1.0.0 beta 3, you can also use "," instead of "=" in FOR.

OUT

51

Function

Output logic levels to the OUT pins.

Format

```
OUT [<terminal>,] <value>
```

Example

```
OUT 1,1
OUT 63
```

123456789012345678912345678 9012

Comment

<terminal> is 1 to 11, 5 and 6 in 0.9.0 or later,
7 to 11 in 1.0.2 beta 3 and 1.1.0 or later,
8 to 11 correspond to 1.1 beta 5 and later.
In IchigoJam shipped from around May 2014 - January 2015
OUT5 is written as EX3, OUT6 as EX4.
The LED will also be OUT7. (OUT 7, n is the same as LED n)
In IchigoJam BASIC RPi, OUT7 and LED are handled separately.

When <terminal> is not attached, <value> is 0 to 2047,
OUT1 to 6 and LED and IN1 to 4 are
It corresponds to 1, 2, 4, 8, 16, 32, 64, 128, 256, 512, 1024.

When <terminal> is attached, <value> is 0 or 1.
This setting is valid after 0.8.7. The output voltage is 3.3V.
For 1.1 beta 5 and later, <value> becomes -1 to 1,
When <value> is set to -1 Allocate OUT1 to 4 and IN(5) to (8) and ANA(5) to (8).

From 1.1 beta 6 onwards, IN(10) to (11) also correspond to OUT5 to 6.

1.2 From beta 43, when <value> is set to -2, use pull-up input.
It can be set except for IN3 (OUT10).

In IchigoJam RPi, IN (OUT10) also corresponds to pull-up input.

For 1.1 beta 5 and later, IN1 - 4 correspond to OUT8 - 11.

When using it, specify OUT as the target, such as OUT 8, 1.
OUT10 (IN3) is treated as open drain and there is no resistance.
Please use it by pulling up externally when using.

SkyBerryJAM also has a lamp equivalent to OUT1 to 4 terminals.
These lamps change with the OUT command.
Since the LED corresponds to OUT7, for example OUT 1 + 2 + 4 + 8 + 64
Or OUT 1,1 : OUT2,1 : OUT 3,1 : OUT 4,1 : OUT 7,1,
you can turn on all the lamps.

PEEK

52

Function

Get the current value from virtual address location.

Format

PEEK (<number>)

Example

PEEK(#E0*8)

1234567890123456789123456789012

Comment

#000 to #FFF are effective for <number>.
Between #000 and #100F is valid from 1.1 beta 8.

The following are valid in 0.9.9 – RC 6 or later.
Some characters such as character patterns are effective before that.

Change of character pattern corresponds to <address> #000 to #7FF,
Pattern code 0 (#00) to 255 (#FF) patterns can be acquired.
Eight addresses (+0 to +7) are composed of one character from character code number ×8.
It is entered from the top in the order of the youngest of the numbers.
Unlike POKE, all character codes can be acquired.

Arrays and variables correspond to <address> #800 to #8FF.

The most significant bit is the sign bit in the order of the lower 8 bits and the upper 8 bits.

Continuing from #800 to #801 for [0], #802 to #803 for [1] ..., up to [101] for #8CA to #8CB,

Subsequently, variables #8CC to #8CD are from variable A and become variable Z for #8FE to #8FF.

The screen display character corresponds to #900 to #BFF.

#900 is a character code with abscissa 0 and ordinate 0, followed by the character code,

It is up to the abscissa (x-axis) 31 and the ordinate (y-axis) 23 of #BFF.

The program corresponds to #C00 - #1001.

(From #1000 to #1001 from 1.1 beta 8)

Line number 2 address (lower 8 bits, upper 8 bits in this order), after the character number 1 address (including the end code).

The program itself excluding the line number is entered.

The end code is #00 and 1 byte.

If the next address is odd, add another byte #00.

As a result, the line number is always made to be an even number.

When it becomes 1023 to 1024 bytes, the end code enters #1000 to #1001, 1.1 beta 8 and later can detect this.

The key status corresponds to #1002 to #100F. (From 1.1 beta 8)

#1002 Arrow key · Space key status (1 ← · 2 → · 4 ↑ · 8 ↓ ·)
#1003 Number in key buffer
#1004 ~ Key Buffer

When entering machine language, machine language body,
you can get via PEEK a value acquisition result of USR.
Because IchigoJam PC has a different CPU, USR can not be used.

PLAY

53

Function

To play a sound.

Format

```
PLAY "<MML>"
PLAY <start address>
```

Example

```
PLAY "CDE2CDE2"
PLAY A
```

123456789012345678912345678912

Comment

<MML> = Music Macro Language, is as follows:
Up to 128 characters are allowed up to 1.2 beta 2.

A to G note (corresponding to CDEFGAB)
 + Increase semitone (no high pitch effect)
 # Increase by a half tone (1.2 beta 10 more than high pitch effect is not available)
 - Lower a semitone (no high pitch effect)
R rest

 <length> length (1 · 2 · 4 · 8 · 16 · 32. attached after the note C4D1)

. Half the length (attach after the length E2.)
T <tempo> The initial value is 120. Can also be specified with command TEMPO
L <length> default length. The initial value is 4
O <Octave> 1 to 5 (1.0.0 beta 6 to 8 are 0 to 9),
    The initial value is 4 (3 from 1.2 beta 10)
> (Up to 1.2 beta 9)
<(From 1.2 beta 10)
    Increase by 1 octave
<(Up to 1.2 beta 9)
> (From 1.2 beta 10)
    Lower one octave
$ Repeat MML from here on
N <height> Sounds sound. Common to BEEP

The range is from O1C to O5D. (O1C to O5B from 1.2 beta 10)

Since 1.0.0 beta 5, the buffer has been changed,
In direct mode, playback stops when another command is executed.

Connect a piezoelectric sounder to SOUND and GND to produce sound.
In IchigoJam shipped from May 2014 - January 2015,
the SOUND terminal is described as EX2.
For 1.0.0 beta 6 ~ 8, SOUND and EX 2 terminal and OUT 5 / EX 3 terminal were exchanged,
It was returned in 1.0.0 beta9.

From <beta 3> <start address> can be specified.
This is usually a variable, starting address where MML is put in the string.
Also, the limit of the number of characters from 1.2 beta 3 is gone.

IchigoJam BASIC RPi supports 1.2b17RPi and 1.2.5RPi.
Connect the piezoelectric sounder to the physical PIN 29 and to GND.

POKE 54

Function

Put a value at a location in the virtual address area.

Format

```
POKE <address>, <value> [, <value>] [, ...]
```

Example

```
POKE #E0*8,#FF
POKE #E0*8,#FF,#81,#81,#81,#81,#81,#81,#FF
POKE #700,#01,#1C,#00,#20,#40,#18,#01,#39,#FC,#D1,#70,#47
```

```
123456789012345678912345678 9012
```

Comment

<address> is #700 to #FFF, <value> is #00 to #FF,
<value> can be described consecutively.

The following are valid in 0.9.9 - RC 6 or later.
Some characters such as character patterns are effective before that.

Changes to the character pattern correspond with <address> #700 to #7FF.
(You can not change #000 to #6FF)
You can change character codes 224 (#E0) to 255 (#FF).
Eight addresses (+0 to +7) are composed of one character from character code number ×8.

They are entered from the top in the order of the latest of the numbers.

Arrays and variables correspond to <address> #800 to #8FF.
The most significant bit is the sign bit in the order of the lower 8 bits and the upper 8 bits.
Continuing from #800 to #801 for [0], #802 to #803 for [1] ..., up to [101] for #8CA to #8CB,
Subsequently, variables #8CC to #8CD are from variable A and become variable Z for #8FE to #8FF.

The screen display characters correspond to #900 to #BFF.
#900 is a character code with abscissa 0 and ordinate 0, followed by the character code,
It is up to the abscissa (x-axis) 31 and the ordinate (y-axis) 23 of #BFF.
IchigoJam BIG is #900 to #9BF and has abscissa 15 and ordinate 11 of #9BF.
Unlike PRINT, when updating this area, it will not be output serially.

The program corresponds to #C00 - #FFF.
Line number 2 address (lower 8 bits, upper 8 bits in this order),
After character number 1 address (including end code 0)
the program itself excluding the line number is entered.
0 is entered as the end code of the line.
If the next address is an odd number, enter 1 byte 0 again,
The line number should be an even number.

Machine language is also supported. It can be executed via USR.
The area can be changed like character pattern, array / variable.
Please use the free space part of the program.

As the IchigoJam PC uses a different CPU, the function USR can not be used,
And POKE is not available.

PRINT

55

Function

It displays on the screen. It outputs as well to serial.

Format

```
PRINT "<character string> ["] [;]
PRINT <number> [;]
? "<string> ["] [;]
? <number> [;]
```

Example

```
PRINT "HI!";
? "HI!
LOCATE 0,-1:?"on UART ";:?A
```

123456789012345678912345678 9012

Comment

If it is a character string, enclose it with ".
If it becomes a new line after that, the trailing " can be omitted.
If you attach a semicolon (;) at the end, it is possible to concatenate.
If you do not add a ";" on the bottom line, the display will scroll up one line.
It is possible to use "?" Instead of PRINT.

When serial connection is made, it is output as it is.

When the vertical position is set to -1 in LOCATE (for example, LOCATE 0, -1), it does not output to screen output, but only serial output.

PWM   56

Function

Perform pulse width modulation on the selected OUT terminal.

Format

```
PWM <terminal>, <pulse width> [, <pulse period>]
```

Example

```
PWM 2,10
PWM 2,10,20
PWM 2,4800,-9600 (1.2 beta 13 ~)
```

123456789012345678912345678 9012

Comment

<terminal> corresponds to OUT2 ~ 5 with terminal pins 2 ~ 5. The other side uses GND or OUT1.

<pulse width> sets the time when it becomes HIGH in 0.01 millisecond increments (100 = 1 millisecond).

Values from 100 to 200 are used for the servo.

<pulse period> specifies the pulse period in units of 0.01 milliseconds.

The default is 2000, so 20 milliseconds.

1.2 beta When it is minus value from 13, <pulse width> and <pulse cycle> It becomes 480 times faster.

PWM 2,10*480, 20*-480 and PWM 10,20 have the same operation.

IchigoJam BASIC RPi only supports terminals 3 and 4.

REM 57

Function

Ignore the contents described after this statement to the end of the line.

Format

```
REM <comment>
'<comment>
```

Example

```
REM START
'START

123456789012345678912345678 9012
```

Comment

All characters after REM are treated as comments and are ignored to the end of the line.
Even if you enter a command separated by ":" it is treated as a comment and is ignored.
This can be used for debugging, for commands not to be temporarily executed.
It is possible to put REM in the head and ignore it.

You can use ' instead of REM.

RENUM   58

Function

To change the line numbers of the program.

Format

```
RENUM [<start line number>]
RENUM [<start line number>] [, <increment>]
```

Example

```
RENUM
RENUM 100
```

123456789012345678912345678 9012

[Comment

When RENUM is executed, it changes to 10 step line numbers.
<start line number> can be used from 0.9.8 RC 11. If omitted, it is from 10.
The line numbers of GOTO and GOSUB are not changed. They must be changed manually.
0.9.8 From RC 6, lines that need to change line numbers are displayed.

1.1 <increment> is added and the increment value of the line number can be set.
The default is 10. Up to 1.1 beta 10 was fixed at 10.

RESET   59

Function

Restart IchigoJam BASIC. (Software reset)

Format

```
RESET
```

Example

```
RESET
123456789012345678912345678912
```

Comment

This function is equivalent to GND - RESET connection or power-off and power-on again.

RETURN

60

Function

Used to return to the calling program from the last GOSUB.

Format

RETURN
RTN

Example

RETURN
RTN

123456789012345678912345678 9012

Comment

After executing the subroutine called via GOSUB,
RETURN is used to return to the original routine.
If RETURN is executed while GOSUB is not being used, an error occurs.

Abbreviated RTN can be used from 1.1.0 beta 1.

RND   61

Function

Generate a random number.

Format

```
RND (<number>)
```

Example

```
RND (6)
```

123456789012345678912345678 9012

Comment

Returns an integer number greater than or equal to 0 and less than <number>.
For example, if <number> is 6, it returns a random number
(0, 1, 2, 3, 4, 5,)
Initialization can be performed using SRND.

RTN - short for RETURN 62

Function

It will return from last GOSUB.

Format

RTN
RETURN

Example

RTN
RETURN

123456789012345678912345678 9012

Comment

After executing the subroutine via GOSUB,
it is used to return to the original main routine.
If RETURN is executed while GOSUB is not being used, an error occurs.

Abbreviated RTN can be used from 1.1.0 beta 1.

RTN is an alias for RETURN.

RUN 63

Function

Start executing the program from the lowest line number.

Format

RUN

[Example]

RUN

123456789012345678912345678 9012

Comment

Execute the program from the first line.

SAVE 64

Function

Save the program.

Format

```
SAVE [<program number>]
```

Example

```
SAVE 1

123456789012345678912345678 9012
```

Comment

The value of <program number> is 0 to 3 and
it is 100 to 226 or 100 to 227.

It corresponds to 0 to 3 from 1.0.0 beta. (Including IchigoJam PC)
The IoT compatible version is only 1.2b59 IoT and the program is only saved to 0.
When omitted, it is 0 until 0.9.8 RC,
It becomes the program number which accessed latest from 0.9.9 RC.
If you omit it immediately after startup, save will go to 0.

From 0.9.3, a save to external EEPROM. 100 to 226 is valid.
Values from 100 to 227 are valid from 0.9.9 RC.

Valid values vary depending on the EEPROM capacity and the EEPROM wiring.
A program number that can not be used returns File Error.

EEPROM can be used with 32K to 1024Kbits,
When using 1024K bits with 0.9.9 RC then the contents can be completely used except for some EEPROM.
For the 32K to 256Kbits it is the same program.
Please be careful to refer to multiple program numbers.
Because there is a difference in the access of I2C for EEPROM of 16Kbit or less, and it can not be used for saving programs.

IchigoJam BASIC RPi is 0 to 3 and 100 to 227 to 1.2b15 RPi · 1.2.4 RPi.
(micro) Save to the file in the files directory of the SD memory card.
It became 0 to 227 from 1.2b16 RPi and 1.2.5 RPi.
0 to 99 are SD memory cards, 100 to 227 are saved in EEPROM.

IchigoJam PC supports 0 to 3 number saving, but it stays in RAM memory and so it is deleted when the power is turned off.

If you want to delete a saved program, it is done via a SAVE <program number> with a NEW and empty program,
So the program of <program number> can be deleted.

Technically, virtual address areas #C00 to #FFF are copied to the main body / EEPROM area.
For this reason, POKE puts in a vacant area of the program (area behind #FFF) with a value
Via SAVE you can also save this value.
After LOAD and LRUN this value can be retrieved with PEEK.

SCR   65

Function

Get the character of coordinates.

Format

```
VPEEK (<abscissa>, <ordinate>)
SCR (<abscissa>, <ordinate>)
VPEEK ([<abscissa>, <ordinate>])  (0.9.0 ~)
SCR ([<horizontal coordinate>, <vertical coordinate>])  (0.9.0 ~)
```

Example

```
VPEEK(0,0)
SCR()
```

1234567890123456789123456789012

Comment

<abscissa> = <x-axis>
<ordinate> = <y-axis>
<vertical coordinate> = <ordinate> = <y-axis>

The coordinates that can be normally acquired are
up to 0.9.8 RC, <abscissa> is 0 to 35, <ordinate> is 0 to 26.

After 0.9.9 RC, <abscissa> is 0 to 31, and <vertical coordinate> is 0 to 23.
The character code at this location is returned as the value.
SCR is an alias for VPEEK.

From 0.9.0, when the coordinates are omitted,

Then get the character at the current cursor position.

SCROLL

Function

Scroll up, down, left and right.

Format

```
SCROLL <direction>
```

Example

```
SCROLL 2

123456789012345678912345678901 2
```

Comment

Scroll one character up, down, left, or right in one direction.
<direction> is as follows.
UP, DOWN, LEFT, RIGHT are supported from 1.0.0 beta 7.

```
0 · UP    = up
1 · RIGHT = right
2 · DOWN  = down
3 · LEFT  = left
```

SLEEP   67

Function

Pauses program execution.

Format

```
SLEEP
```

Example

```
SLEEP
123456789012345678912345678 9012
```

Comment

Execution of the program is stopped by SLEEP, and it waits for button input.
When IchigoJam's button is pressed, it returns to the IchigoJam startup state.
Since it will be a motion that completely reset (power OFF → ON)
Hold down the button a few seconds
(Starting from 1.0.0 beta and keeping it pressed even after BEEP sounds)

You can start program number 0.
Beginning with 1.0.0 beta 16, always start the program number 0 after returning from SLEEP.
In this case, startup display is no longer performed.

IchigoJam BASIC RPi does not have this function implemented.

SOUND   68

Function

Detects the presence or absence of a running sound output.

Format

```
SOUND()
```

Example

```
50 IF SOUND() GOTO 50

123456789012345678912345678 9012
```

[Comment

It returns 1 if a sound is currently generated, and a 0 if there is no output.

In direct mode, playback stops by this command execution,
And if it was being played then you can detect it via SOUND.

SRND 69

Function

Initialize the random number seed for RND.

Format

```
SRND <value>
```

Example

```
SRND 0

123456789012345678912345678 9012
```

Comment

Change the random number table that RND refers to by <value>.

STOP 70

Function

Stop program execution.

Format

```
STOP
```

Example

```
STOP
123456789012345678912345678 9012
```

Comment

When the STOP is executed, program execution is stopped. You can run it again from this line you stopped via CONT.

STR$ 71

Function

Returns a string.

[Command format]

```
STR$(<start address> [, <length>])
```

Example

```
PRINT STR$(A)
PRINT STR$(M+5,3)
```

123456789012345678912345678 9012

Comment

Outputs a character string whose start address is defined via <start address>.
Normally it is only a variable, but you can specify it like A+1.
A="ABCDE":PRINT STR$(A+2) will output CDE.
You can specify the number of characters to output with <length>. If omitted, all of them are output.
A="ABCDE":PRINT STR$(A+2,2) will output CD.

SWITCH   72

[Function]

Switch between video output and liquid crystal display output.

Format

SWITCH [<mode>]

Example

SWITCH
SWITCH 1

123456789012345678912345678 9012

Comment

The values of <mode> are as follows:
The default is 0.

   0 Video output
   1 Liquid crystal display output

When <mode> is omitted, then switch between video output and liquid crystal output.
SWITCH is assigned to function key F10.

In DakeJacket compatible version (has D in version)
SWITCH has been deleted in 1.2b 56 IoT.
However, in IoT compatible version it can be used from 1.2b59 IoT.
IchigoJam BASIC RPi does have this function implemented.

TEMPO

Function

Specify the tempo of PLAY.

Format

```
TEMPO <tempo>
```

Example

```
TEMPO 120

123456789012345678912345678 9012
```

Comment

It is the same function as T for PLAY in <MML>.

IF~THEN(~ELSE)   74

Function

Perform conditional branching.

Format

```
IF <condition value> [THEN] <command>
IF <condition value> [THEN] <command> [ELSE <command>]
```

Example

```
IF A=B THEN LED 1
IF A=B LED 1

123456789012345678912345678912
```

Comment

Execute <command> when <condition value> is correct (when it is not 0).

<condition value> corresponds to multiple conditions by AND and OR from 0.9.4.
Up to 0.9.3 substitute with * and + or & and |. (Condition values before and after are delimited by () pair)

Multiple executions of <command> are possible, then separated by ":".
It will cause the command within that line to continue.

Although THEN can be omitted, the line number immediately after THEN becomes an error.

To skip, please use GOTO line number or THEN GOTO line number.

When <condition value> is not met (in case of 0)
you can execute the command immediately after ELSE. It is optional.
ELSE is supported from 0.9.4.

TICK   75

Function

Returns the elapsed time of the TICK counter.

Format

TICK()

Example

PRINT TICK()

123456789012345678912345678901 2

Comment

When IchigoJam is running, it is always iincrementing the count state,
In the NTSC version, it advances in 1/60 seconds steps. And it is a count of about 60 for 1 second.
In the PAL version, it advances by 1 in about 1/50 of a second. So, for about 1 second it is a count of 50.
The PAL version is supported from 1.2.2. It is NTSC version in Japan.
When VIDEO is set to 0, counting is not performed while it is not displayed.
Next to 16383 (#3FFF), it returns to 0.
1.0.0 beta 10 is changed to 0 after 32767 (#7FFF).
Even in clock down mode of VIDEO from 1.1 beta 4 please be aware that it will be affected.

When it is set to VIDEO 1,10 then the values of TICK() becomes 10 seconds instead 60 seconds.

Up to 0.9.7 is reset to 0 by RUN etc.,
It is also possible to reset to 0 using CLT.
Since 0.9.8 is no longer reset by RUN etc.
To reset, you need to use CLT.

UART 76

Function

Set the output of the serial interface

Format

```
UART <mode 1>
UART <mode 1> [, <mode 2>]  (1.1 beta 13 ~)
```

Example

```
UART 0
UART 1,0
```

123456789012345678912345678 9012

Comment

<mode 1> is as follows:
The startup default is 2
for IchigoJam BASIC and IchigoJam BASIC RPi,
It is 0 in IchigoJam PC.

   0 Do not output
   1 Print only PRINT. Line feed code LF
      It will be close to the behavior before 1.0.
      In 1.1 beta 6 it works the same as UART 3.
      1.1 beta 7 has been restored to the previous specification.

2 Outputs for PRINT, LOCATE, CLS, SCROLL.
   Line feed code LF Output is the following code according to the command.
   LOCATE → CHR$(#15,#20+<X coordinate>,#20+<Y coordinate>)
   CLS → CHR$(#13,#0C)
   SCROLL → CHR$(#15,<cursor direction>)
      <cursor direction> is #1C=left, #1D=right, #1E=up, #1F=down
3 Output only PRINT. Line feed code CR+LF, with buffer.
   It is more effective than 1.1 beta 7.

For Devices that control according to serial signals (such as PanCake) and In applications,
it is necessary to execute UART and change the sending from IchigoJam.

It corresponds to <mode 2> from 1.1 beta13.
It is as follows:
The startup default is 1
for IchigoJam BASIC and IchigoJam BASIC Rpi.
It is 0 in IchigoJam PC.

   0 Ignore sending from serial
   1 Normal operation (Valid when numerical values other than described are specified)
   2 Ignore the ESC key
      Thus, when the ESC code is transmitted from the serial
      it prevents the program from being stopped.
   4 Convert CR to LF (1.2 beta 41 ~)

USR   77

Function

Call machine language program.

Format

```
USR(<address>, <value>)
USR(<address>, <value>, <character pattern>)
   (1.1 beta 6 ~)
USR(<address>, <value>, <character pattern>, <character pattern
top>)
(1.1 beta 10 ~)
```

Example

```
? USR(#700,100)

123456789012345678912345678 9012
```

Comment

Execute machine language from <address> address.
<value> can be acquired from the first parameter (machine language to R0 register).
Since the value of the R0 register is returned as the return value, it is possible to use it with LET or PRINT.
In case of making with C you can get it with the first parameter and return it with return.

From 1.1 beta 6 as the second parameter (R1 register)

Pass the start pointer of virtual memory (corresponding address of #000 part).
However, this value can not be changed; Virtual address of <character pattern>
The #000 to #6FF part actually does not exist. # 700 or later is valid.

From 1.1 beta 10 as the third parameter (R2 register)
Pass the top address of the <character pattern> (corresponding address of #000 part).
Here, #000 to #6FF is effective.

Up to 1.1 beta 5

```
int16_t sample (int16_t val) {
  /* Main */
  return val;
}
```

1.1 beta 6 or later

```
int16_t sample (int16_t val, char * mem) {
  /* Main */
  return val;
}
```

Where to place machine language programs?
· #700 to #7FF Changing character pattern
· #800 to #8FF array and variable
· #C00 to #FFF program body (1.1 beta 10 ~)
These you can use.

Characters and arrays / variables after CHR$(224)
Please be careful not to destroy the program itself.

IchigoJam BASIC and IchigoJam BASIC RPi (Raspberry Pi)
There are differences depending on the CPUs used.

IchigoJ amd RPi A/A+/B/B+/0/0+W support Thumb
RPi 2B+/3/3+ support Thumb 2

IchigoJam PC does have this command implemented. It returns Syntax Error.

VER

78

Function

Get the version number of the IchigoJam software.

Format

VER ( )

Example

? VER ( )

123456789012345678912345678 9012

Comment

The return version number is as follows. (As of August 2018)
IchigoJam PC and IchigoJam ap and
IchigoJam Web by WebAssembly and DakeJacket version,
it becomes the IchigoJam BASIC version number of the derivation source.
IchigoJam BASIC RPi is a unique version number.
An error occurs before 0.9.3.

      94 = 0.9.4
      95 = 0.9.5
      96 = 0.9.6
      97 = 0.9.7 (official version)
      98 = 0.9.8 RC 1 to 12, RC number can not be
              confirmed

99 = 0.9.9 RC 1 to 12, RC number can not be confirmed
10001 to 10016 = 1.0.0 beta 1 to 16, the last 2 digits are beta number
10017 = 1.0.0 (Private on the net · Included in some products)
10100 = 1.0.1 (official version)
10212 = 1.0.2 beta 12, the last two digits are beta numbers *
10201 to 10211 = 1.0.2 beta 1 to 11, the last two digits are beta numbers *
11001 = 1.1.0 beta 1
11002 = 1.1.0 beta 2 (unpublished on the net) or 1.1 beta 2
11003 to 11017 = 1.1 beta 3 to 17, the last two digits are beta number
11015 = Ichigo Jam PC 0.1 beta 12
11018 = 1.1.0 (not disclosed on the net)
11019 = 1.1.1 (official version)
12001 to 12025 = 1.2 beta 1 to 25, the last 2 digits are beta number
12025 = 1.2.0 (official version, same as 1.2 beta 25)
12026 = 1.2.0 IchigoJam BIG
12027 = 1.2 beta 27
12027 = 1.2.1 (official version, same as 1.2 beta 27)
12230 to 12247 = 1.2 beta 30 to 47, the last two digits are beta numbers
12247 = 1.2.2 (official version, same as 1.2 beta 47)
12348 = 1.2 beta 48
12348 = 1.2.3 (official version, same as 1.2 beta 48)
12011 = IchigoJam BASIC RPi 1.2.0 RPi (official version) ★

12314 = Ichigo Jam BASIC RPi 1.2 b 14 RPi
12314 = IchigoJam BASIC RPi 1.2.3 RPi
   (official version, same as 1.2 b 14 RPi)
12415 = IchigoJam BASIC RPi 1.2 b 15 RPi
12415 = IchigoJam BASIC RPi 1.2.4 RPi (the same
   as the official version, 1.2 b 15 RPi) ★
12516 to 12519 = IchigoJam BASIC RPi 1.2 b 16 RP i to 1.2
   b 19 RPi, the last two digits
   are beta number
12449 to 12461 = 1.2b 49 to 61, the last two digits are
   beta numbers. Common for IoT
   12456 = 1.2.4 D (Dake Jacket version)
   12457 = 1.2.4 D 2 (Dake Jacket version)
12520 to 12522 = IchigoJam BASIC RPi 1.2 b 20 RPi to 1.2
   b 22 RPi, the last two digits are beta
   number
   12522 = IchigoJam BASIC RPi 1.2.5 RPi (official
      version, same as 1.2 b 22 RPi) ★
   12623 = IchigoJam BASIC RPi 1.2.6 RPi
      (official version) ★

※ 1.0.2 beta 12 is the successor from 1.0.1,
   It is taken over from 1.0.1 beta 11 to 1.1.0 beta 1,
      later 1.1 beta.
   However, 1.0.2 is not officially released, 1.1.0 is
      officially released.
★ IchigoJam BASIC RPi is a derivation from 1.2.3 up to
   1.2.4 RPi.
   1.2.3 In addition to RPi, please note that 1.2.0 RPi
      and 1.2.4 RPi also comes from 1.2.3.
   1.2.5 RPi to 1.2.6 RPi contains specifications of
      1.2b56 and 1.1.2b57 IoT.

VIDEO 79

Function

Display / hide the screen output.

Format

```
VIDEO <display>                   (~ 1.1 beta 3 · RPi)
VIDEO <display> [, <clock down>]  (1.1 beta 4 ~)
```

Example

```
VIDEO 0
VIDEO 0,10
```

123456789012345678912345678 9012

Comment

In <display>, set display / non-display of output to the video terminal.
0 is hidden,
1 is displayed.
In non-display, in order to turn processing speed in line drawing to other processing, it will be able to do process and line drawing faster.
VIDEO 0 is valid even in direct mode.
Please be aware that any input will be completely invisible.
You can go back to display with VIDEO 1.

You can do clock down from 1.1 beta4.
By lowering the speed it is possible to use lower power consumption.

1 <clock down> is 48 MHz by default,
2 is 1/2, 3 is 1/3, and the maximum value is 255.
The value of TICK() ·and WAIT is also affected.
If set to 10, it will be about 10 seconds for TICK() ·
WAIT to be 60.

1.2 Since beta 9, 2 has been added to <display>.
Highlight the monochrome of the video output. The display area is flipped. Since this effect has the effect of changing the entire screen, there is a possibility that it may have a bad influence if it is heavily used.
Especially please do not make flashing continuously.

1.2 beta 3 - 8 has been added to <display> from 34-5.
It corresponds to enlarged display. It is as follows.

- · 0 hidden
  - · 1 to 2   1 times (32 × 24)
  - · 3 to 4   2 times (16 × 12)
  - · 5 to 6   4 times (8 × 6)
  - · 7 to 8   8 times (4 × 3)

  - · 1 · 3 · 5 · 7   black background · white character
  - · 2 · 4 · 6 · 8   white background · black character

The same display as IchigoJam BIG released at 1.2.0 is possible with VIDEO 3.

If the screen size changes, clear the screen.
The screen is not cleared if there is no change in display / inversion and display, no change.

In IchigoJam BASIC of LPC 1114 operation, VIDEO 5 to 8
may become a display that can shake on the left and the right.
When continuous output is done with PRINT,

when there is no change in the display, this can be reduced by not letting PRINT.
Moreover, this does not occur in IchigoJam BASIC RPi.

IchigoJam BASIC RPi supports 1.2b17RPi and 1.2.5RPi.
<display> only, the <clock down> is not implemented.

VPEEK   80

Function

Get the character at the coordinates of the display.

Format

```
VPEEK (<abscissa>, <ordinate>)
SCR (<abscissa>, <ordinate>)
VPEEK ([<abscissa>, <ordinate>]) (0.9.0 ~)
SCR ([<horizontal coordinate>, <vertical coordinate>])
    (0.9.0 ~)
```

Example

```
VPEEK(0,0)
SCR()
```

1234567890123456789123456789012

Comment

<abscissa> = <x-axis>
<ordinate> = <y-axis>
<vertical coordinate> = <ordinate> = <y-axis>

The coordinates that can be normally acquired are
Up to 0.9.8 RC, <abscissa> is 0 to 35, <ordinate> is 0 to 26,

After 0.9.9 RC, <abscissa> is 0 to 31, and <vertical coordinate> is 0 to 23.

The character code is returned for the value at this location.

SCR can be used instead of VPEEK.

From 0.9.0, when the coordinates are omitted,
Get the character at the current cursor position.

WAIT   81

Function

Stop processing for a certain period of time.

Format

```
WAIT <time>
WAIT <time> [, <dive>]  (1.2 beta 7 ~)
```

Example

```
WAIT 60
WAIT 60,0

123456789012345678912345678 9012
```

Comment

Stop processing for the specified time.
<time> is about 1 second for 60 in NTSC version, and about 1 second for 50 in PAL version.
The range is from 1 to 32767. (
NTSC version in Japan, PAL version is from 1.2.2)
added since 1.1 beta4,
It is affected by VIDEO clock down.
With VIDEO 0,10 then WAIT 60 will be about 10 seconds.

By putting 0 in <dive> from 1.2 beta 7,

Suppresses current consumption and keeps memory and pin states in function deep sleep.
IchigoJam BASIC RPi does not support Deep Sleep.

By putting a negative value into <time> from 1.2 beta 12,
it will wait for scanning line × -1 minute.
-262 is 1 frame (1/69 seconds).

Commands

82

Function

Commands most used outside a program.

Format

?ANA([<value>]) - print the voltages of IN2 and BTN
?ASC("<character>") - print character code
BEEP [<period>, <length>] - tone generator
?BIN$(<value> [, <number of digits>]) - print binary
BPS [<communication speed>] [, <I2C communication speed>]
?BTN(0) - print button key (if it exists) being pressed
?CHR$(<character code>) - print character
CLV   - clear keyboard buffer and key state
CLO   - clear input / output state
CLP   - clear PCG font changes
CLS   - clear visible screen area
CLT   - clear TICK() timer
CLV   - clear variables
CONT   - continue program execution
COPY<destination>, <source>, <length> - memcopy
?DEC$(<value> [, <number of digits>]) - print decimal
?FILE() - print last accessed file number
FILES - list files
FILES [<Maximum number of files>]
?FREE() - print amount of free memory
GOTO<line> - continue program execution at <line>
HELP   - print memory map
?HEX$(<value> [, <number of digits>]) - print hexidecimal
LET <variable> = <value> - assign variable

LIST [[<start line>,] <end line>] - program
LOAD   - load 0 or last program, clear variables
LOCATE <x-axis>, <y-axis> - set position of cursor
LOCATE <x-axis>, -1 - set position of serial cursor
LRUN   - load and run 0 or last program
LRUN [<program number>] [, <line number>]
NEW   - erase the program in memory, clear variables
OUT [<terminal>,] <value> - output to the OUT terminal
PLAY <start address> - play <MML> in memory or variable
POKE <address>, <value> [, <value>] [, ...] - memory ※
PRINT "<character string> ["] [;] - to screen or serial
RENUM [<start line number>] [, <increment>] - program
RESET - soft reset
RETURN - return from last GOSUB and execute program
RUN   - start program execution at 1st line
SAVE   - save or delete 0 or last program
SLEEP - pause program execution, button start ★
SRND 0 - reset random number generator seed
SWITCH - toggle video/LCD
?TICK() - print ticks since last reset
VIDEO 0 - turn video output off
VIDEO 1 - turn video output on

Constants 83

[Function]

Constant Variables.

Format

SCROLL() - UP, DOWN, LEFT, RIGHT are supported from 1.0.0 beta 7.
  0 · UP = up
  1 · RIGHT = right
  2 · DOWN = down
  3 · LEFT = left

BTN() - 0 SPACE UP DOWN LEFT RIGHT
  0 · 0 = button key (if it exists)
  0 · UP = up arrow
  2 · DOWN = down arrow
  1 · RIGHT = right arrow
  3 · LEFT = left arrow
  3 · SPACE = space bar

Functions 84

Function

Quick list of Functions.

Format

ABS(<value>) – return absolute value
ANA([<value>]) – return the voltages of IN2 and BTN
ASC("<character>") – character code
ASC(<variable>) – character code
BTN([<value>]) – UP DOWN LEFT RIGHT SPACE 0
BIN$(<value> [, <number of digits>]) – binary string
CHR$(<character code>) – one character
CHR$(<character code>, <character code>, ...) – string
DEC$(<value> [, <number of digits>]) – decimal string
FILE() – last accessed file number
FREE() – amount of free memory
HEX$(<value> [, <number of digits>]) – hexidecimal string
I2CR(<I2C address>, [<command start address>, <command
　　length>,> <read memory starting address>, <read
　　memory length>) – read from I2C device ※
I2CW(<I2C address>, [<command start address>, <command
　　length>,] <write memory start address>, <write
　　memory length>) – write to I2C device ※
IN([<terminal>]) – input from terminal
INKEY() – character of key being pressed
IoT.IN() – receive data from sakura.io module ※
LANG() – returns the language type
LEN(<start address>) – length of the string

LINE() - line number being executed
PEEK(<number>) - memory peek
RND(<number>) - generate a random number, see SRND n
SCR([<x-axis>, <y-axis>]) - screen, alias for VPEEK
SOUND() - presence or absence of sound output
STR$(<start address> [, <length>]) - return a string
TICK() - ticks since last reset
USR(<address>, <value>) - execute machine code ※
VER() - get version number
VPEEK([<x-axis>, <y-axis>]) - video peek, empty=cursor

Statements 85

Function

Quick list of Statements.

Format

'<comment> - short form REMark
BEEP [<period>, <length>] - tone generator
BPS [<communication speed>] [, <I2C communication speed>]
CLEAR - clear variables, alias CLV
  CLV - clear keyboard buffer and key state
  CLO - clear input / output state
  CLP - clear PCG font changes
  CLS - clear visible screen area
  CLT - clear TICK() timer
  CLV - clear variables
COPY <destination>, <source>, <length> - memcopy
ELSE - optional in IF~THEN~ELSE
END - end program execution
FILES [[<start file number>,] <end file number>] - list
FILES [<Maximum number of files>]
FOR <variable>=<number> TO <number> [STEP <step>] - NEXT
GOTO <line> - jump to <line>
GOSUB <line> - branch to subroutine at <line>
GSB <line> - gosub, alias for GOSUB
IF (eval) THEN (true) ELSE (false)
INPUT ["<string>",] <variable> - user input
IoT.OUT <value> - send data to sakura.io module ※
KBD <keyboard array> - only on RPi
LC - alias for LOCATE

LED <number> - same as OUT7, turn on / off the lamp
LET <variable> = <value> - assign variable
LET [<array>], <number>, number, ... - number array
LET [<array>], "<string>", "<string>", ... - string array
LOAD [<program number>] - load n, 0 or last program
LOCATE <x-axis>, <y-axis> - set position of cursor
LOCATE <x-axis>, -1 - set position of serial cursor
LRUN [<program number>] [, <line number>]
NEXT - end for FOR~NEXT loop
OUT [<terminal>,] <value> - output to the OUT terminal
PLAY "<MML>" - Music Macro Language
PLAY <start address> - play <MML> in memory or variable
POKE <address>, <value> [, <value>] [, ...] - memory ※
PRINT "<character string> ["] [;] - to screen or serial
PWM <terminal>, <pulse width> [, <pulse period>] ※
REM <comment> - program line comment
RESET - soft reset
RETURN - return from last GOSUB
RTN - return, alias RETURN
SAVE [<program number>] - save the program
SCROLL <direction> - scroll screen
SLEEP - pause program execution, button start ★
SRND <value> - seed random number generator
STOP - stop program execution
SWITCH [<mode>] - switch video/LCD ★※
TEMPO <tempo> - T of PLAY in <MML>
THEN - optional in IF~THEN~ELSE, can be GOTO
UART <mode 1> [, <mode 2>] - serial output mode
VIDEO <display> [, <clock down>] - change video
WAIT <time> [, <dive>] - seconds or scanlines (-time)

Testing the Instructions

Understanding the instructions works best, if there is a piece of code executing them and to have the relevant hardware that might be required.
The LED, the button and the buzzer are on the board already, but they are fixed to their pins.
The minimum I found, was an LED and a piece of wire plus at least 2 resistors to do more (plus a microservo).
A better and electrically more correct way would be not to connect the LED directly to any outputs pins directly, so the next step would be:

A little breadboard
Connecting wires
One or 2 LED-resistor pairs
2 additional resistor to test the analog input
A microservo to show the PWM working
An LDR – Light Dependant Resistor.

Some additional components I prepared for testing code for testing

There is a microbit kit with 2 servo-based motors - to be tested as well.

Links Part 1

BALLOON a.k.a. Fu-sen. <balloonakafusen@gmail.com>

Juergen,
I am pleased that you are interested in IchigoJam.
IchigoJam developers want to extend IchigoJam to the world. Your support will be appreciated.

PCN, which sells products from IchigoJam, sells a book in English. This includes the BASIC reference:
https://pcn-en.stores.jp/items/573e657099c3cd56f8005643

You may wish that these IchigoJam products will be sold on Amazon, including that book.
(Several products are on sale at Amazon.co.jp:
https://www.amazon.co.jp/s?k=ichigojam&ref=nb_sb_noss_1)

If you are concerned about using PCN informations, please contact PCN. They can also respond in English:
https://pcn.club/products/index.en.html
--> There is an email address in the page footer

Also, if you think you need, please contact the jig.jp subsidiary B-inc, which owns IchigoJam's license.
They can also respond in English:
https://ichigojam.net/index-en.html
--> support - inquiry

The official reference is a supplement to the archive:
https://ichigojam.net/IchigoJam-en.html
This is included with IchigoJam BASIC running on LPC1114.
but some features are different in IchigoJam BASIC RPi, so they are not included.

The command reference I have published explicitly specifies the CC BY 4.0 license:
https://github.com/fu-sen/IchigoJam-BASIC
https://github.com/paulwratt/IchigoJam-BASIC-english
(CC BY license=クリエイティブ・コモンズ 表示 4.0 国際 ライセンス)
Therefore, you can make it a book freely without getting my reply.

Paul's translation for the English version uses automatic translation, so some terms seem difficult to understand.

And this reference reflects the latest beta version from time to time.
The latest beta version of IchigoJam BASIC is "1.3.2b24" while returning this email.
(It will be "1.4.0" in the near future)
However, these are all commands that are not described in the IchigoJam attached reference.
If you are aiming for the IchigoJam BASIC RPi, this base version of "1.2.6RPi" is "1.2.4".
I think we will be adding commands within a few months.

If you are using Facebook, posting to this group is useful:
https://www.facebook.com/groups/ichigojamfan/

Ias I now have your contact, so I'll get in touch with IchigoJam developers and PCN members and spread the message.

BALLOON a.k.a. Fu-sen. (Keiichi SHIGA)

https://15jamrecipe.jimdo.com/basic/ichigojam-pc/
https://fukuno.jig.jp/1256
https://15jamrecipe.jimdo.com/basic/ichigojam-pc/
https://15jamrecipe.jimdo.com/basic/ichigojam-web/
https://fukuno.jig.jp/app/IchigoJam/

Some Program Code Preparation

Our screen is 32 characters wide. If you want nicely formatted code, make sure it fits in the screen width. This line of numbers shows how wide the text is:

12345678901234567891234567890012

Here in this chapter, our summary of code is shown in the Examples of each instruction and numbered from line number xx0 to xx9.
And then the instructions are shown in line number steps of 10, so each instruction example can have 9 lines of code. If there are more lines needed than these 9, then there is a GOTO xxx to a narea where the extra code is, and from there a GOTO back to the next instruction.
For this the lines 1000 onwards are prepared for now.

As well the maximum of letters per line can be added
12345678901234567891234567890012

But here with REM in front of it and a bit shortened to ensure we stay within the 32 character width if we want.

You can write your example code in here if you wish.

```
1  REM  789012345678912345678900012
2  REM
3  REM
4  REM
5     REM  LIST OF INSTRUCTIONS
10    REM  ABS
20    REM  ANA
30    REM  ASC
40    REM  BEEP
50    REM  BIN$
60    REM  BPS
70    REM  BTN
```

```
80  REM  CHR$
90  REM  CLEAR
100 REM  CLK
110 REM  CLO
120 REM  CLP
130 REM  CLS
140 REM  CLT
150 REM  CLV
160 REM  CONT
170 REM  COPY
180 REM  DEC$
190 REM  ELSE
200 REM  END
210 REM  FILE
220 REM  FILES
230 REM  FOR
240 REM  FREE
250 REM  GOSUB
260 REM  GOTO
270 REM  GSB
280 REM  HELP
290 REM  HEX$
300 REM  I2CR
310 REM  I2CW
320 REM  IF
330 REM  IN
340 REM  INKEY
350 REM  INPUT
360 REM  IOT.IN
370 REM  IOT.OUT
380 REM  KBD
390 REM  LANG
400 REM  LC
410 REM  LED
420 REM  LEN
430 REM  LET
440 REM  LINE
450 REM  LIST
460 REM  LOAD
```

```
470 REM    LOCATE
480 REM    LRUN
490 REM    NEW
500 REM    NEXT
510 REM    OUT
520 REM    PEEK
530 REM    PLAY
540 REM    POKE
550 REM    PRINT
560 REM    PWM
570 REM    REM
580 REM    RENUM
590 REM    RESET
600 REM    RETURN
610 REM    RND
620 REM    RTN
630 REM    RUN
640 REM    SAVE
650 REM    SCR
660 REM    SCROLL
670 REM    SLEEP
680 REM    SOUND
690 REM    SRND
700 REM    STOP
710 REM    STR$
720 REM    SWITCH
730 REM    TEMPO
740 REM    THEN
750 REM    TICK
760 REM    UART
770 REM    USR
780 REM    VER
790 REM    VIDEO
800 REM    VPEEK
810 REM    WAIT
1000 REM
1010 REM
1020 REM
1030 REM
```

```
1040 REM
1050 REM
1060 REM
1070 REM
1080 REM
1090 REM
1100 REM
1110 REM
1120 REM
1130 REM
1140 REM
```

The following block of REMs is the same as on the cover page and shows the pin functions of the 28 LPC1114 pins, the 5 pin interface connector and the pinout of the USBtoTTL converter I used.

```
123456789012345678912345678 9012

2000 REM IchigoJam Chip    / CN5
2001 REM  1 VID1   KBD1 28 1 5V
2002 REM  2 VIDEO2 SCL  27 2 SDA
2003 REM  3 IN1    KBD2 26 3 SCL
2004 REM  4 IN2   SOUND 25 4 3V
2005 REM  5 IN3     ISP 24 5 GND
2006 REM  6 IN4   RESET 23
2007 REM  7 VCC     GND 22 / USB
2008 REM  8 GND     VCC 21 1 GND
2009 REM  9 OUT1     X1 20 2 RXD
2010 REM 10 OUT2     X2 19 3 TXD
2011 REM 11 OUT3   OUT5 18 4 5V
2012 REM 12 OUT4   OUT6 17 5 3V3
2013 REM 13 BTN     TXD 16 6 DTR
2014 REM 14 LED     RXD 15
```

And the same a bit smaller – I used it for a business card sizeed helper: Pins on one side, the List of Instructions on the other side.

```
123456789012345678912345678901 2
 
2000 REM IchigoJam Chip   / CN5
2001 REM  1 VID1  KBD1 28 1 5V
2002 REM  2 VIDEO2 SCL 27 2 SDA
2003 REM  3 IN1   KBD2 26 3 SCL
2004 REM  4 IN2   SOUND 25 4 3V
2005 REM  5 IN3    ISP 24 5 GND
2006 REM  6 IN4   RESET 23      
2007 REM  7 VCC    GND 22 / USB
2008 REM  8 GND    VCC 21 1 GND
2009 REM  9 OUT1    X1 20 2 RXD
2010 REM 10 OUT2    X2 19 3 TXD
2011 REM 11 OUT3   OUT5 18 4 5V
2012 REM 12 OUT4   OUT6 17 5 3V3
2013 REM 13 BTN    TXD 16 6 DTR
2014 REM 14 LED    RXD 15      
```

LIST OF ICHIGOJAM BASIC INSTRUCTIONS

ABS	ANA	ASC	BEEP	BIN$	BPS	BTN
CHR$	CLEAR	CLK	CLO	CLP	CLS	CLT
CLV	CONT	COPY	DEC$	ELSE	END	FILE
FILES	FOR	FREE	GOSUB	GOTO	GSB	HELP
HEX$	I2CR	I2CW	IF	IN	INKEY	INPUT
IOT.IN	IOT.OUT	KBD	LANG	LC	LED	LEN
LET	LINE	LIST	LOAD	LOCATE	LRUN	NEW
NEXT	OUT	PEEK	PLAY	POKE	PRINT	PWM
REM	RENUM	RESET	RETURN	RND	RTN	RUN
SAVE	SCR	SCROLL	SLEEP	SOUND	SRND	STOP
STR$	SWITCH	TEMPO	THEN	TICK	UART	USR
VER	VIDEO	VPEEK	WAIT	version	1 June	2019

It is always good to have an overview of the available IchigoJam BASIC instructions handy.

And here as picture – just a screen print to copy out and one side of a business card.

And on the other side could be the pinouts of the board, the interface of the USBtoTTL board.

As I work my way through this project, I will see what else I have to have available quickly.

```
LIST  OF  ICHIGOJAM  BASIC  INSTRUCTIONS¶
¶
ABS····ANA······ASC····BEEP····BIN$····BPS····BTN··¶
CHR$····CLEAR····CLK····CLO····CLP····CLS····CLT··¶
CLV····CONT····COPY····DEC$····ELSE····END····FILE·¶
FILES···FOR······FREE····GOSUB····GOTO····GSB····HELP·¶
HEX$····I2CR····I2CW····IF······IN······INKEY··INPUT¶
IOT.IN·IOT.OUT·KBD····LANG····LC······LED····LEN··¶
LET····LINE····LIST····LOAD····LOCATE··LRUN···NEW··¶
NEXT····OUT······PEEK····PLAY····POKE····PRINT·PWM··¶
REM·····RENUM····RESET··RETURN·RND·····RTN····RUN··¶
SAVE····SCR······SCROLL·SLEEP··SOUND···SRND···STOP·¶
STR$····SWITCH··TEMPO··THEN····TICK····UART···USR··¶
VER·····VIDEO····VPEEK··WAIT····version·1·June·2019·¶
¶
```

Some Links

https://www.facebook.com/groups/ichigojamfan/ English Facebook group and examples doc

https://github.com/IchigoJam/IchigoJam.github.io?fbclid=IwAR0aBhjFUBbmyEb74NJELWhlhsFkmDsw4xQg1vaYouC98tmXTMz9DgZZuXk different versions

https://yrm006.wordpress.com/2018/05/24/ichigonquest-for-microbit-beta-has-been-released-you-can-do-programming-without-laptop-and-tablet-microbit%e7%89%88-ichigonquest-%ce%b2%e7%89%88%e3%83%aa%e3%83%aa%e3%83%bc%e3%82%b9%ef%bc%81/?fbclid=IwAR3pYIxPYuBZt2RPRV1L_IYbgzz_zbJvC5Yip-dR2_lrpmZlPU6ejmxAhU microbit

https://www.facebook.com/notes/ichigojam-fanen/exchange-programs-with-sd-card-of-ichigojam-basic-rpi/849312032117954/ sd card to transfer programs

https://15jamrecipe.jimdo.com/pancake/%E4%BD%BF%E3%81%84%E6%96%B9/ Ichigo Pancake

https://fukuno.jig.jp/english?off=130&id=english Fukuno Taisuke's one day creation-create every day

http://forum.fritzing.org/t/ichigojam-t-as-a-kids-pc/4997/5 on Fritzing

To be expanded as new interesting links are found.

OLIMEX LPC1114 Board

Still to be tested, but it had worked well in the past with the Forth MPE VFX ARM LITE, and here via serial interface. MPE's AIDE includes the terminal software.

The question here is as well if all of the IOs can be addressed, or only the ones defined in IchigoJam.

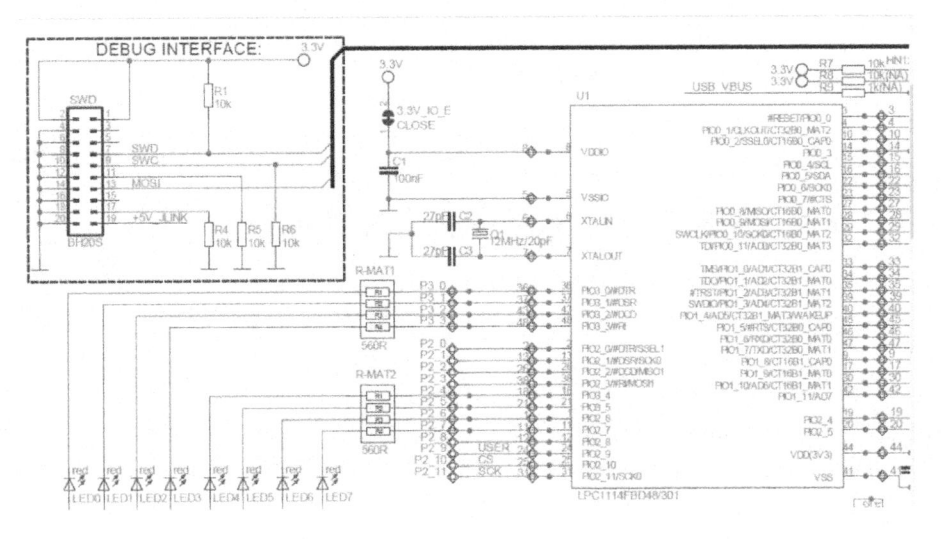

149

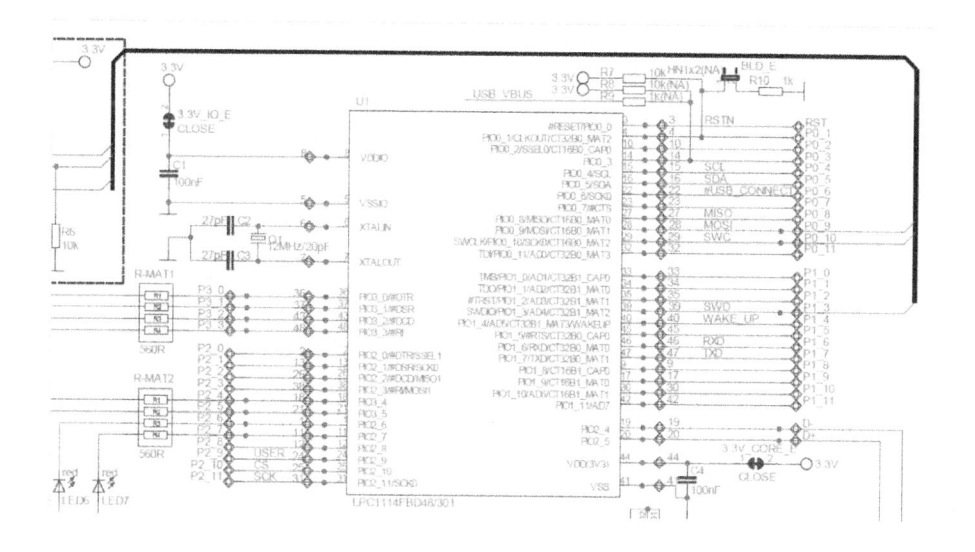

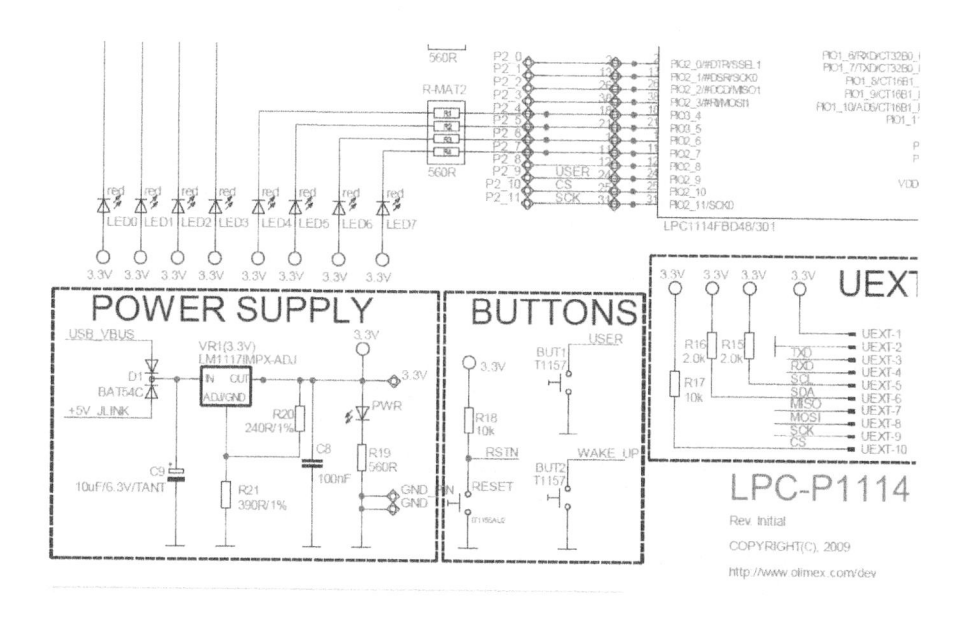

This product is available as well outside Japan, and at low cost, so might be an easy start to try out the IchigeJam BASIC for the ones who want to solder. Especially if you only want to use it via the serial interface.

2 resistors for video is probably the minimum needed for the normal option.

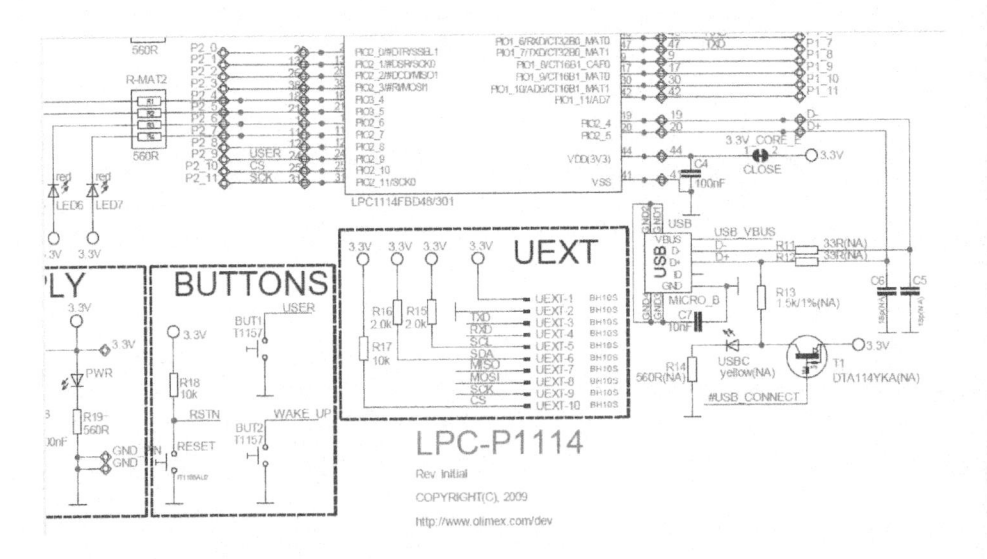

[Schematic diagram of LPC-P1114 board by Olimex, Rev Initial, COPYRIGHT(C) 2009, http://www.olimex.com/dev]

See www.olimex.com and search for LPC1114

LPC1114 Pin Functions

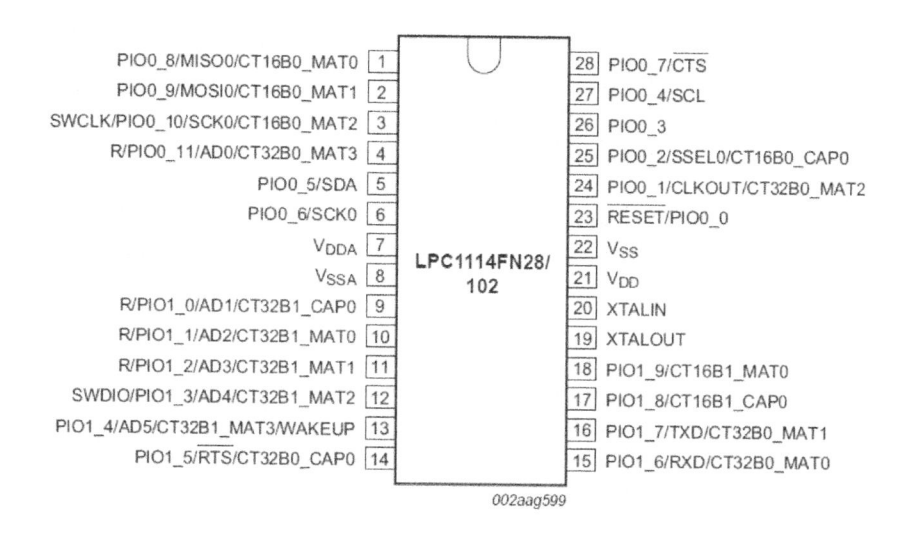

```
PIO0_8/MISO0/CT16B0_MAT0    [1]              [28] PIO0_7/CTS
PIO0_9/MOSI0/CT16B0_MAT1    [2]              [27] PIO0_4/SCL
SWCLK/PIO0_10/SCK0/CT16B0_MAT2  [3]          [26] PIO0_3
R/PIO0_11/AD0/CT32B0_MAT3   [4]              [25] PIO0_2/SSEL0/CT16B0_CAP0
PIO0_5/SDA                  [5]              [24] PIO0_1/CLKOUT/CT32B0_MAT2
PIO0_6/SCK0                 [6]              [23] RESET/PIO0_0
VDDA                        [7]  LPC1114FN28/ [22] VSS
VSSA                        [8]      102     [21] VDD
R/PIO1_0/AD1/CT32B1_CAP0    [9]              [20] XTALIN
R/PIO1_1/AD2/CT32B1_MAT0    [10]             [19] XTALOUT
R/PIO1_2/AD3/CT32B1_MAT1    [11]             [18] PIO1_9/CT16B1_MAT0
SWDIO/PIO1_3/AD4/CT32B1_MAT2 [12]            [17] PIO1_8/CT16B1_CAP0
PIO1_4/AD5/CT32B1_MAT3/WAKEUP [13]           [16] PIO1_7/TXD/CT32B0_MAT1
PIO1_5/RTS/CT32B0_CAP0      [14]             [15] PIO1_6/RXD/CT32B0_MAT0
```

002aag599

IchigoJam – BASIC Reference version 1.3

v10_A5 Juergen Pintaske April 2019

Keyboard Usage

Operation	Description
Key	To type in a character/number
Shift	The key to type in a symbol character with keys and to do capitals
KANA / CRTL+Shift	Switch the mode between Alphabet or Katakana (Rome-ji)
Enter	To execute, and to edit
Shift+Enter	To separate a line
ESC	To stop program execution, to show a list, to show a file list
Cursor Key	To move the cursor
Backspace	To erase a character located left of the cursor
Delete	To erase a character located at the current cursor position
ALT	The key to type in some special characters together with 0-9 / A-K / [] keys
Home, End	To move the cursor to the beginning of the line, to move the cursor to the end of the line
Page Up, Page Down	To move the cursor to the top, to move the cursor to the bottom of the display page
Caps	Switch to capital letters mode, or to the small letter mode
Insert / CTRL+ALT	Switch to Insert Mode, or Overtype Mode
Function Keys	F1:clear screen; F2:LOAD; F3:SAVE; F4:LIST; F5:RUN; F6:?FREE; F7:OUT0; F8:VIDEO1; F9:Files; F10:SWITCH
Button	General functions, or auto execution mode if you push this button when you turn on the board

BASIC Commands
– more complete list and condensed; unfortunately here different numbers, but still helpful.

1	Command	Description	Example
2	LED num	Light up the LED when n is 1, switch off when n = 0	LED 1 or LED 0
3	WAIT num(,num2)	Wait n frames (60 frames =1 sec) (if num2 equals 0 lower powerconsumption mode, if num1 is minus then wait short mode-261, same as WAIT1	WAIT 60
4	:	To connect multiple commands in one line	WAIT 60:LED 1
5	line num commands	Memory commands with line number	10 LED 1
6	linenum	Delete the command in memory which has this line number	10
7	RUN	Execute program in memory (F5) starting at the lowest line number	RUN
8	LIST (line num1 (, line num2)	Show the program in memory[F4} (line num1: show the line,if minus to the line / linenum2:, show to this line, if 0 to the end / ESC to stop)	LIST 10,300
9	GOTO line num	Jump to another program execution line, (it's OK to use variables)	GOTO 10
10	END	End this program	END
11	IF num (THEN) command1 (ELSE command2)	If num does not equal 0, execute command 1; else execute command2; (you can omit THEN / ELSE	IF BTN( ) RND
12	BTN ((num))	Return 1 , if you push the	LED BTN( )

		button, else 0 (num0 (embedded button) / UP/ DOWN/ RIGHT/ LEFT/ SPACE, 0.no num/	
13	NEW	Delete all current programs in memory	NEW
14	PRINT (num or strings)	Write the letter or number to the screen (strings must be surrounded by ("); connect pars via (:I) Abbreviation ?	PRINT "HI"
15	LOCATE x,y,num	Set the position to write to (if y equals -1 no rite mode); show the cursor if num not equals 0. Abbreviation: LC	LOCATE 3,3
16	CLS	Clear the screen	CLS
17	RND	Return a random number between 0 and to num - 1	PRINT RND(6)
18	SAVE	Save the current program in memory (num 0-3, 100-227:optional EEPROM, if omit using number)	SAVE 1
19	LOAD	Load the proram number 0 3 , 100-227: optional EEPROM, if omit using number)	LOAD
20	FILES	Show the file list from num1 to num2 (all if num1 equals 0, ESC to stop)	FILES
21	BEEP (num1(num2))	Sound the BEEP, num1 is period(1-255), num2 length (1/60sec) (you can omit num1 and num2 – to connect the sounder on SOUND(EX2)-GND	BEEP
22	PLAY	Play the music specified mml as MML(Music Macro Language) just play to stop the music – to connect the	PLAY "CDE2CDE2"

		sounder on SOUND(EX2)-GND	
23	TEMPO	Change the temo of playing music	TEMPO 1200
24	x+y	Return x+y	PRINT 1+1
25	x-y	Return x-y	PRINT 2-1
26	x*y	Return x*y	PRINT 7*8
27	x/y	Return Integer of x divided by y	PRINT 9/3
28	x%y	Return Remainder of x divided by y	PRINT 10%3
29	(num)	Return calculate the number in priority	PRINT 1+(1*2)
30	LET var,num	Set the number to 1 letter of alphabet as named memory (variable) (series put to the array) Abbreviation: var=num	LET A.1
31	INPUT strings, var	Set the number to var from keyboard input (you can omit strings and comma)	INPUT "ANS",A
32	TICK()	Return the time count from CLT (count up in 1/60 sec)	PRINT TICK()
33	CLT	Clear the time count	CLT
34	INKEY ()	Return from keyboard or UART (0:no input, #100:0 input from UART	PRINT INKEY()
35	constant	LEFT=28, RIGHT=29, UP=30, DOWN=31, SPACE=32	IF INKEY()=SPACE LED1
36	CHR$(num)	In PRINT, return the letter string specifified in the num (you can set series with comma)	PRINT CHR$(65)
37	ASC("string")	Return the letter code from a string	PRINT ASC("A")
38	SCROLL num	Scroll the screen (0/UP:up, 1/RIGHT:right, 2/DOWN:down, 3/LEFT:left	SCROLL 2

39	SCR ((x,y))	RETURN the letter code located at x,y on the screen (if omit x and y, use current position Alias VPEEK	PRINT SCR(0,0)
40	x=y	Return 1 if x equals y, else 0 (Alias ==)	IF A=B LED 1
41	x<>y	Return 1 if x does not equal y, else 0 Alias!=)	IF A<>B LED 1
42	x<=y	Return 1 if x <=y, else 0	IF A<=B LED 1
43	x<y	Return 1 if x < y, else 0	IF A<B LED 1
44	x>=y	Return 1 if x >= y, else 0	IF A>=B LED 1
45	x>y	Return 1 if x > y, else 0	IF A>B LED 1
46	x AND y	Return 1 if x AND y , else 0 (Alias &&)	IF A=1 AND B=1 LED 1
47	x OR y	Return 1 if x or y, else 0 (Alias:||)	IF A=1 OR B=1 LED 1
48	NOT x	Return 1 if x equals 0, else 0 (Alias:!)	IF NOT A=1 LED 1
49	REM	Do not execute after this command, (comment) Abbreviation ')	REM START
50	FOR var=num1 TO num2 (STEP num3) / NEXT	Set num1 to var, execure the loop to the NEXT until var reaches num2 by step num3 (you can omit STEP, nest limit: 6)	FOR I=0 TO 10:?|
51	IN((num))	Return 1, if wheninput terminalpin is HIGH, else 0 (num0-11, IN0/1/4/9 pullup, IN5-8, 10-11; if switched. IN0,9:button,; you can get all states when you omit num)	LET A,IN(1)
52	ANA((num))	Return the value 0-1023 specified voltage of input terminal (2:IN2, 5-8:IN5-8 (OUT1-4), 0,9:BTN, 0:omitted)	?ANA()
53	OUT num1(,num2)	Output num2 to the output	OUT 1,1

		pin specified via num1 (num1:OUT1-11; you can set all states when you omit num2, if num2 equals -1, the outputswitch into input pin; if num2 equals -2, the output pin switches to input with Pull-up)	
54	PWM num1,num2(,num3)	Output num2(0.01msec) length pulse in num3(if omit 2000) period to the output pin specified num1 (num1:OUT2-5,OUT2-4 same period)	PWM 2,100
55			
56	MML Music Macro Language		
57	Command	Description	Example
58	Tone	Tone specified by a letter from (C D E F G A B)(R is rest,space just skip)	CDER FG
59	Tone num	Sound a tone in by num specified length (with . half stretch the length)	C4 E2 D1 F32
60	Tone+	Sound a tone in a half tone higher	C+ D+
61	Tone-	Sound a tone in a half tone lower	D- E-
62	Tn	Set the tempo (you can changewith TEMPO command) initial value 120	T96CDE
63	Ln	Default length (1,2,3,4,8,16,32) initial value:4	CL8DC
64	On	Set the Octave from O1C to O5B, initil value :3	O3CO2C
65	<	Octave up	C<C<C
66	>	Octave down	C>C>C
67	$	Repeat play after this mark	C$DE

68	Nn	Sound a tone specified by 1-255 (same as BEEP command)	N10N5
69	'	End of Music	C'DE
70			
71	Senior (Advanced) Commands		
72	Command	Description	Example
73	CLV	Clear (set to zero) variables and array variables Alias:CLEAR	CLV
74	CLK	Clear key buffer and key status	CLK
75	CLO	Initialize the input and output pins	CLO
76	ABS(num)	Return the absolute value	?ABS(-2)
77	(num)	Array variables (from (0) to (101) 102 series variables you can set in series using LET(0),1,2,3	(3)=1
78	GOSUB linenum / RETURN	Move execution to linenum and execute after this command when RETURN Abbreviation:GSB/RTN (nest limit : 30)	GOSUB 100
79	DEC$(num1) (,num2)	In PRINT, return strings from num1 with beam specified num2(you can omit num2)	?DEC$(99,3)
80	#hexsum	Return the number specified in hexadecimal	#FF
81	HEX$(num1(,num2))	In PRINT, return hexadecimal strings from num1 with beam specified num2 (you can omit num2)	?HEX$(255,2)
82	'binnum	Return the number specified in binary number	'0101
83	BIN$(num1(,num2))	In PRINT, return binary	?BIN$(255,8)

		number strings from num1 with beam specified in num2 (you can omit num2)	
84	x & y	Return x logical AND y (bit calculation)	?3&1
85	X | y	Return x logical OR y (bit calculation)	?3|1
86	X^y	Return x logical EXCLUSIVE OR y (bit calculation)	?A^1
87	x>>y	Return x Shifted Down by y bits (bit calculation)	?A>>1
88	x<<y	Return x Shifted Up by y bits (bit calculation)	?A<<1
89	~x	Return bit INVERTED x (bit calculation)	?~A
90	STOP	Stop the program	STOP
91	CONT	Continue the same line or stop line	CONT
92	SOUND()	Return 1 if sound playing, else return 0	?SOUND()
93	FREE()	Return free memory for program (up to 1024 bytes)	?FREE()
94	VER()	Return the version number of IchigoJam BASIC	?VER()
95	LANG()	Return the language number of IchigoJam BASIC (1:Japanese, 2:Mongol, 3:Vietnamese)	?LANG()
96	RENUM (num1 (,num2))	Renumber the line number of program from num1 step num2 (num):10, num2:10 if omit, you may have to change manually line number specified in GOTO/GOSUB	RENUM
97	LRUN (num)	LOAD num and RUN	LRUN 1
98	FILE()	Return the number of the last used FILE	?FILE()
99	LINE()	Return the line number of	?LINE()

			last execution	
100	SRND()		Initialize the seed of random/td>	SRND()
101	HELP		Display the memory map	HELP
102	PEEK(num)		Read one byte number from the memory location specified in num2	?PEEK(#700)
103	POKE num1,num2		Write 1 byte as specified in num1 to memory location specified in num2	POKE #700,#FF
104	COPY num1, num2, num3		Memory copy from num1 to num2, length specified via num3 (if num3 is minus, copy direction is inverted)	COPY#900,0,256
105	CLP		Initialize the Character Pattern Memory (~700-#7FF)	CLP
106	"strings"		Return the address of strings in the memory	A="ABC"
107	STR$(num1(,num2))		In PRINT, return strings from address specified num1 (num2 length you can omit)	PRINT STR$(A)
108	LEN("strings")		Return the length of strings	PRINT LEN("ABC")
109	@label		Put in front of the line, you can use as destination line number (type GOTO, @loop)	@LOOP
110	VIDEO num1(,num2)		Switch the video signal to enabled or disabled(num0:0 diabled / 1 enable F8/ 2invert black and white / greater than bigmode, num2:when num0 is 0, clock down mode clock=1 / num2)	VIDEO 0
111	RESET		Reboot the IchigoJam	RESET
112	SLEEP		Sleep the IchigoJam (after push button, execute and	SLEEP

		run file 0)	
113	UART num1(,num2)	Set UART mode (num) 0:off 1:with PRINT/LC/CLS/SCROLL, 3:with PRINT and entercode is \r\n, echo back input with +4, initial value:2) (num2- 0:UART recv off 1:on initial value:1)	UART 0
114	BPS num1(,num2)	Num1:setUART speed (0:115200-1:57600- 2:38400-and so on to 9600 .. if -100 or less, -100 times speed. For example, if it is - 2604, it is 260400 initial value :0); numn2: set I2C speed (unit kHz, 0:400kHz, value:0)	BPS 9600
115	OK (num)	Show OK and error messages (1:messages on of omit; 2: messages off)	OK 2
116	I2CR (num1,num2, num3, num4, num5)	Num4: addressof return data, num5:length of return data (if command is 1 byte you can omit num2, if no commands num2 num2 /num3 you can omit)	R=I2CR(#50,#700, 2, #702,2)
117	I2CW (num1,num2,num2, num4, num5)	Write to I2C device num1:I2C address, num2: address of command, num3 length of command, num4:address of return data, num5:length of return data (you can omit num4/num5, if command is 1 byte you can omit num2)	RI2CW(#50,#700, 2, #702,2)
118	IoT.IN()	Get a received number from the sakura.io module	R=IoTIN()
119	IoT.OUT num	Send a number to send via	IoT OUT 100

		the sakura.io module as the 0 channel	
120	SWITCH (num)	Switch the video output TV or LCD; (0:TV 1:LCD)	SWITCH
121	USR(address,num)	Call program written in machine code (warning: freezes of IchioJam with a little mistake)	A=usr(#700,0)

Operator Precedence

Level	Operator	Description
1	( )	brackets
2	- ~	NOT	minus, bitwise NOT, logical NOT
3	* / % MOD << >> & ^	multiplicative, shift, bitwise AND,
4	+ - !	plus, minus, bitwise OR
5	= != < > <= >=	comparison
6	AND	logical AND
7	OR	logical OR

CC BY https://ichigojam.net/ https://creativecommons.org/licenses/by/4.0/ v10

Current List of Example Programs

This part here is work in progress and the code parts to be tested on the board and the RPI first.

A decision has been taken to rather make this available with a partial list of example codes rather than to wait until all has been done and tested.
As this is spare time work, it could take another 6 months. More important is to make the reference available for others to use, for feedback and to then modify the final and print version.

There will be two Example Program Versions:

The first one shows more explanations/text and needs more memory.
Try it via FREE ( ).
The second version will use less description and as well the shorter way for instructions.And the difference in memory usage will be seen.

My Thanks goes to Eiichi-Inohara San who has kindly sent most of the examples here for me to test and include into this documentation.

I am waiting as well for a 512kBit EEPROM board, as I want to test the instructions related to it.

```
123456789012345678912345678012

1  REM DEMO Programs
2  REM MY REFERENCE shows
3  REM numbers - multiplied
4  REM by 10 gives the example
5  REM starting point just
6  REM type GOTO xx to execute

10 REM ABS - return absolute
12 REM a 23 -> 23, -23 -> 23
13 LET A= 23
14 ABS (A)
15 PRINT A
16 LET B = -23
17 ABS (B)
18 PRINT B
19 END

20 REM ANA returns analog value
22 REM 2:IN2, 5-8:IN5-8(OUT1-4
23 REM ), 0,9:BTN, 0:omitted
24 REM ANA(5) uses IN5=OUT1
25 LET C=ANA(5)
26 PRINT (C)
27 END

30 REM ASC return decimal value

32 LET D=(A)
33 PRINT A
33 N=ASC("A")
34 ?N
35 END

40 REM BEEP  make a sound
```

```
41 FOR A=0 TO 30
42 F=RND(30)+1:'Frequency
43 L=5          :'Length
44 BEEP F,L
45 WAIT 10
46 NEXT
47 END


50 REM BIN$
51 N=#50AF
52 ?BIN$(N)
53 ?BIN$(N,16)
54 END


60  REM


70  REM BTN
71 CLS:?"Push Cursor or [Esc]"
72 X=15:Y=11
73 LOCATE X,Y
74 X=X+BTN(RIGHT)*(X<31)
75 X=X-BTN(LEFT)*(X>0)
76 Y=Y+BTN(DOWN)*(Y<22)
77 Y=Y-BTN(UP)*(Y>1)
78 ?CHR$(RND(32)+224);
79 WAIT 5:GOTO 73


80  REM CHR$
81 FOR A=32 TO 255
82 ?CHR$(A);
83 NEXT
84 END


90  REM CLEAR
91 A=5
92 ?"A=";A
93 CLEAR
94 ?"A=";A
95 END
```

```
100  REM

110  REM

120  REM CLP
121  FOR A=224 TO 255
122  ?CHR$(A);
123  NEXT
124  CLP
125  FOR A=#700 TO #7FF
126  POKE A,PEEK(A)^#FF
127  NEXT
128  WAIT 40
129  GOTO 124

130  REM CLS
131  FOR A=0 TO 300
132  POKE #900+RND(768),RND(256)
133  NEXT
134  CLS
135  GOTO 131

140  REM

150  REM

160  REM CONT
161  [0]="*_":[1]="_*"
162  CLS
163  ?"Press [Esc]key to Stop"
164  ?"Type 'CONT' and press [Enter]key"
165  LC0,3:?STR$([RND(2)]):CONT

170  REM COPY
171  COPY #900,RND(#FFF),768
172  WAIT 20
173  GOTO171
```

```
180 REM DEC$
181 CLS
182 FOR A=0 TO 20
183 ?DEC$(RND(1200),5)
184 NEXT
185 END


190   REM ELSE
191 A=RND(2)
192 IF A=0 ?"A" ELSE ?"B"
193 WAIT 10
194 GOTO 191


200   REM END
201 ?"START"
202 END
203 ?"END"


210   REM FILE
211 ?"Current file No.:";FILE()
212 END


220   REM FILES
221 ?"Files in Ichigojam"
222 FILES
223 END


230   REM FOR
231 CLS
232 FOR A=0 TO 255
233 POKE #900+A,A
234 NEXT
235 END


240   REM FREE
241 ?"Free area for Program"
242 ?FREE();" Bytes"


250   REM GOSUB
```

```
251 FOR A=1 TO 10
252 GOSUB @DOUBLE
253 NEXT
254 END
255 @DOUBLE
256 ?A*2
257 RETURN


260   REM GOTO
261 ?"Hello, World! ";
262 GOTO 261


270   REM GSB
271 GOTO 250


280   REM HELP
281 HELP
282 END


290   REM HEX$
291 A=#ABC
292 ?HEX$(A)
293 ?HEX$(A,4)
294 END


300   REM


310   REM


320   REM IF
321 A=RND(100)
322 ?"Counting Game"
323 INPUT Q
324 IF A=Q ?"Correct !!":END
325 IF A<Q ?"Less"
326 IF A>Q ?"More"
327 GOTO 323


330   REM
```

```
340  REM INKEY
341 ?"Press any key"
342 A=INKEY()
343 IF !A GOTO 342
344 FOR Y=0 TO 7
345 FOR X=7 TO 0 STEP -1
346 ?CHR$(PEEK(A*8+Y)>>X&1);
347 NEXT:?
348 NEXT
349 END

350  REM INPUT
351 ?"Input any number (1<n<30)"
352 INPUT A
353 FOR B=1 TO A
354 FOR C=1 TO B
355 ?"*";
356 NEXT
357 ?
358 NEXT
359 END


360  REM

370  REM

380  REM

390  REM

400  REM LC
401 GOTO 470

410  REM LED
411 ?"[SPACE]key to turn on the LED"
412 A=BTN(SPACE)
413 LED A
414 GOTO 412
```

```
420  REM LEN
421  A="ABCDEFG"
422  ?"Length of '";STR$(A);
423  ?"' is ";LEN(A)
424  END

430  REM LET
431  LET A,10:?A
432  A=20:?A
433  [0]=30:?[0]
434  LET[1],40,50,60
435  ?[0],[1],[2],[3]
436  END

440  REM LINE
441  ?"Current running line:";
442  ?LINE()
443  END

450  REM LIST
451  LIST

460  REM LOAD
461  A=FILE()
462  LOAD A

470  REM LOCATE
471  CLS
472  X=RND(32)
473  Y=RND(21)+1
474  ?"LOCATE";X;",";Y
475  LOCATE X,Y
476  ?"*";
477  WAIT 50
478  GOTO 471

480   REM
```

490   REM

500   REM

End of the current examples and end of this preliminary version of My Reference.
I hope you enjoy it.

Feedback please to epldfpga@aol.com.

14 July 2019 v11

Fukuno -San's latest list version 1.4 that I have seen

and a conversion list to my numbers used here to compare. Showing the new instructions added in 1.4.

From: https://fukuno.jig.jp/app/csv/ichigojam-cmd.html#lang=en
I gave them numbers as they appear and cross referenced to see what has not been covered in this book yet.

	1.3	ALPHA	1.4	Fukuno List VERSION 1.4
1	X	33	78	ABS
2	X	34	72	ANA
3	X	35	55	AND
4	X	36	44	ASC
5	X	37	23	BEEP
6	X	38	88	BIN
7	X	39	120	BPS
8	X	40	12	BTN
9	X	41	43	CHR$
10	X	42	75	CLEAR
11	X	43	76	CLK
12	X	44	77	CLO
13	X	45	111	CLP
14	X	46	18	CLS
15	X	47	36	CLT
16	X	48	74	CLV
17	X	49	98	CONT
18	X	50	110	COPY
19	X	52	84	DEC
20	X	55	11	ELSE
21	X	56	8	END
22	X	57	22	FILES

23	X	58	104	**FILES**
24	X	59	63	**FOR**
25	X	60	100	**FREE**
26	X	61	80	**GOSUB**
27	X	62	7	**GOTO**
28	X	63	81	**GSB**
29	X	64	107	**HELP**
30	X	65	86	**HEX**
31	X	66	122	**I2CR**
32	X	67	123	**I2CW**
33	X	68	9	**IF**
34	X	69	71	**IN**
35	X	70	37	**INKEY**
36	X	71	34	**INPUT**
37	X	72	124	**IOT.IN**
38	X	73	125	**IOT.OUT**
39	X	74	17	**LC**
40	X	75	1	**LED**
41	X	76	38	**LEFT**
42	X	77	114	**LEN**
43	X	78	32	**LET**
44	X	79	105	**LINE**
45	X	80	6	**LIST**
46	X	81	21	**LOAD**
47	X	82	16	**LOCATE**
48	X	83	103	**LRUN**
49	X	84	13	**NEW**
50	X	85	66	**NEXT**
51	X	86	59	**NOT**
52	X	87	121	**OK**
53	X	88	57	**OR**
54	X	89	70	**OUT**
55	X	90	108	**PEEK**
56	X	91	24	**PLAY**
57	X	93	109	**POKE**

58	X	95	14	**PRINT**
59	X	96	73	**PWM**
60	X	97	61	**REM**
61	X	98	102	**RENUM**
62	X	99	117	**RESET**
63	X	100	82	**RETURN**
64	X	102	19	**RND**
65	X	103	83	**RTN**
66	X	104	5	**RUN**
67	X	105	20	**SAVE**
68	X	106	46	**SCR**
69	X	107	45	**SCROLL**
70	X	109	118	**SLEEP**
71	X	110	99	**SOUND**
72	X	112	106	**SRND**
73	X	114	97	**STOP**
74	X	115	113	**STR**
75	X	116	127	**SWITCH**
76	X	117	25	**TEMPO**
77	X	118	10	**THEN**
78	X	119	35	**TICK**
79	X	121	119	**UART**
80	X	123	128	**USR**
81	X	124	101	**VER**
82	X	125	116	**VIDEO**
83	X	126	47	**VPEEK**
84	X	127	2	**WAIT**
		1	4	**1**
		2	62	'
		3	87	'
		4	27	-
		5	26	+
		6	60	!
		7	49	!=
		8	112	"

9	85	#
10	30	%
11	89	&
12	56	&&
13	31	(
14	28	*
15	29	/
16	3	:
17	15	?
18	115	@
19	79	[]
20	91	^
21	90	|
22	58	||
23	94	~
24	52	<
25	93	<<
26	51	<=
27	50	<>
28	33	=
29	48	==
30	54	>
31	53	>=
32	92	>>
51	95	**COS**
53	41	**DOWN**
54	68	**DRAW**
92	69	**POINT**
94	67	**POS**
101	39	**RIGHT**
108	96	**SIN**
111	42	**SPACE**
113	65	**STEP**
120	64	**TO**
122	40	**UP**

128   126   **WS.LED**

And here we come to the end of the current preliminary version of this book.

7 July 2019 v11

We hope you enjoyed reading this documentation and it is useful for working with the IchigoJam options.

I like this idea of a minimum solution.

And I have to thank the many people in Japan who helped to make this book possible.

As stated before, I rather wanted to make it available in this not yet finished state; I use it anyway as is.
And it grows as I work on it.

So I thought, why not share it. A more finalized version will be available later with additional information included.

This is a collection of documentation in English, as there is not much for now.

The final version will need a lot more work – and it might take 3 to 6 months, as this is spare time work for me and for fun.